Topics in Neuroscience

Springer

Milano
Berlin
Heidelberg
New York
Barcelona
Hong Kong
London
Paris
Singapore
Tokyo

A. Vincent · G. Martino (Eds)

Autoantibodies in Neurological Diseases

Springer

ANGELA VINCENT
Neurosciences Group
Department of Clinical Neurology
Institute of Molecular Medicine
John Radcliffe Hospital
Oxford, United Kingdom

GIANVITO MARTINO
Neuroimmunology Unit
Department of Neuroscience
San Raffaele Scientific Institute
Milan, Italy

The Editors and Authors wish to thank FARMADES-SCHERING GROUP (Italy) for the support and help in the realization and promotion of this volume

---

Springer-Verlag Italia
a member of BertelsmannSpringer Science+Business Media GmbH

© Springer-Verlag Italia, Milano 2002

ISBN 88-470-0119-6

Library of Congress Cataloging-in-Publication Data: applied for

Typesetting: Photo Life - Ideanet (Milan)
Printing and binding: Staroffset (Cernusco sul Naviglio, Milan)
Cover design: Simona Colombo

Printed in Italy

SPIN: 10831851

*To Enzo
who taught me the art of
reasoning and everything else
worthwhile*
(G. MARTINO)

Foreword

Immune-mediated neurological disorders present an increasing challenge to neurologists.

Recent advances have identified new antibodies, new effector mechanisms and new diagnostic tests; these have greatly enhanced our understanding of disease aetiology and opened up therapeutic opportunities, but a critical understanding of this fast-developing field is not easy to find.

This volume was conceived following a very successful course on neuroimmunology in Bergamo in 2000, and was written mainly by clinicians who are active in the laboratory. It will help to keep practising neurologists and neuroimmunologists abreast of the most recent developments, and to assess new information more critically.

Angela Vincent and Gianvito Martino should be congratulated on master-minding this timely and practical text.

July 2001

J. Newsom-Davis
Professor Emeritus
University of Oxford

Table of Contents

Chapter 10 - Paraneoplastic Neuropathies: a Neuropathological Overview

Chapter 11 - Opsoclonus-Myoclonus Syndrome in Childhood

Chapter 12 - Epilepsy and Autoantibodies

Chapter 13 - Primary and Secondary Vasculitis of the Central Nervous System

Subject Index

List of Contributors

F. Andreetta
Myopathology and Immunology Unit,
Department of Neuromuscular Diseases,
National Neurological Institute
"Carlo Besta", Milan, Italy

A. Annoni
Myopathology and Immunology Unit,
Department of Neuromuscular Diseases,
National Neurological Institute
"Carlo Besta", Milan, Italy

C. Antozzi
Myopathology and Immunology Unit,
Department of Neuromuscular Diseases,
National Neurological Institute
"Carlo Besta", Milan, Italy

F. Baggi
Myopathology and Immunology Unit,
Department of Neuromuscular Diseases,
National Neurological Institute
"Carlo Besta", Milan, Italy

A.P. Batocchi
Institute of Neurology, Catholic University,
Rome, Italy

P. Bernasconi
Myopathology and Immunology Unit,
Department of Neuromuscular Diseases,
National Neurological Institute
"Carlo Besta", Milan, Italy

F. Blaes
Department of Neurology, Justus-Liebig
University, Giessen, Germany

C. Buckley
Neurosciences Group,
Institute of Molecular Medicine, John
Radcliffe Hospital, Oxford, United
Kingdom

L. Clover
Neurosciences Group,
Institute of Molecular Medicine, John
Radcliffe Hospital,Oxford, United
Kingdom

G. Comi
Department of Neuroscience, San Raffaele
Scientific Institute, Milan, Italy

F. Cornelio
Department of Neuromuscular Diseases,
National Neurological Institute
"Carlo Besta", Milan, Italy

G. Cossu
Stem Cell Research Institute, San Raffaele
Scientific Institute, DIBIT, Milan;
Department of Histology and Medical
Embryology, "La Sapienza" University,
Rome, Italy

P. Dalton
Neurosciences Group,
Institute of Molecular Medicine, John
Radcliffe Hospital,Oxford, United
Kingdom

A. Evoli
Institute of Neurology, Catholic University,
Rome, Italy

A.M. Ferrarini
Department of Neurologic and Psychiatric
Sciences, Second Neurological Clinic,
University of Padua, Italy

F. Folli
Unit for Metabolic Diseases and
Department of Medicine, San Raffaele
Scientific Institute, Milan, Italy

B. Giometto
Department of Neurologic and
Psychiatric Sciences, Second
Neurological Clinic, University of Padua,
Italy

T. Granata
Department of Child Neurology,
National Neurological Institute
"Carlo Besta", Milan, Italy

W. Hoch
Max Planck Institute for Developmental
Biology, Tübingen, Germany

J. Honnorat
Neurologie B, Hôpital Neurologique,
Lyon, France

B. Lang
Neurosciences Group,
Institute of Molecular Medicine, John
Radcliffe Hospital, Oxford, United
Kingdom

R. Liguori
Institute of Neurology, University of
Bologna, Italy

A. Manfredi
Department of Medicine, San Raffaele
Scientific Institute, Milan, Italy

R. Mantegazza
Myopathology and Immunology Unit,
Department of Neuromuscular Diseases,
National Neurological Institute
"Carlo Besta", Milan, Italy

V. Martinelli
Department of Neuroscience,
San Raffaele Scientific Institute, Milan,
Italy

G. Martino
Neuroimmunology Unit, Department of
Neuroscience, San Raffaele Scientific
Institute, DIBIT, Milan, Italy

J. McConville
Department of Clinical Neurology,
Neurosciences Group, Institute of
Molecular Medicine, John Radcliffe
Hospital, Oxford, United Kingdom

M. Milani
Myopathology and Immunology Unit,
Department of Neuromuscular Diseases,
National Neurological Institute
"Carlo Besta", Milan, Italy

L. Moiola
Department of Neuroscience,
San Raffaele Scientific Institute, Milan,
Italy

P. Montagna
Institute of Neurology, University of
Bologna, Italy

P. Nicolao
Department of Neurologic and
Psychiatric Sciences, Second
Neurological Clinic, University of Padua,
Italy

L. Passerini
Myopathology and Immunology Unit,
Department of Neuromuscular Diseases,
National Neurological Institute
"Carlo Besta", Milan, Italy

G. Piccolo
Department of Neurology, Fondazione
Mondino, University of Pavia, Italy

A. Quattrini
Department of Neuroscience, San
Raffaele Scientific Institute, Milan, Italy

M.G. Sabbadini
Department of Medicine, San Raffaele
Scientific Institute, Milan, Italy

T. Scaravilli
Department of Neurologic and
Psychiatric Sciences, Second
Neurological Clinic, University of Padua,
Italy

B. Tavolato
Department of Neurologic and
Psychiatric Sciences, Second
Neurological Clinic, University of Padua,
Italy

P.A. Tonali
Institute of Neurology, Catholic
University, Rome, Italy

M. Vianello
Department of Neurologic and
Psychiatric Sciences, Second
Neurological Clinic, University of Padua,
Italy

A. Vincent
Neurosciences Group,
Department of Clinical Neurology,
Institute of Molecular Medicine,
John Radcliffe Hospital, Oxford,
United Kingdom

B. Vitaliani
Department of Neurologic and
Psychiatric Sciences, Second
Neurological Clinic, University of Padua,
Italy

Introduction

Approaches to Understanding Immune-Mediated Neurological Disorders: Measuring and Evaluating Autoantibodies to Neuronal Antigens

G. Martino[1], A. Vincent[2]

There is increasing evidence of the relevance of the immune system to the pathogenesis of a variety of neurological disorders, including neuromuscular junction disorders, immune-mediated demyelinating disorders, and also some form of epilepsy, stroke, Alzheimer's and Parkinsons' diseases, and prion diseases. It is, therefore, particularly important to distinguish those conditions which are caused by dysfunction of the immune system, from those in which immune responses play a purely secondary role. In this chapter, we first provide an abbreviated introduction to neuroimmunology, and then discuss ways of measuring immune responses to neuronal antigens, and of approaches that evaluate their pathogenic roles.

Special Features of Relevance to Immune-Mediated Disorders of the Nervous System

The "Immune Privilege" of the Nervous System

In 1948 Medawar demonstrated that foreign skin transplanted into the brain is tolerated over extended periods of time, thus indicating that the nervous system can be considered an "immuno-priviledged" site. In its early time, the "immune privilege" status of the nervous system was thought to be due to a deficit in immune reactivity and immune surveillance. Nowadays, the immune privilege is known to be the result of active as well as passive processes [1]. For instance, delayed type hypersensitivity is impaired in the anterior chamber of the eye due to aqueous humoral factors [2]. Nevertheless, there are many factors that tend to reduce immune responses in the central nervous system (CNS):

[1] Neuroimmunology Unit, Deptartment of Neuroscience, San Raffaele Scientific Institute, DIBIT, Via Olgettina 58, 20132 Milan, Italy. e-mail: g.martino@hsr.it

[2] Neurosciences Group, Department of Clinical Neurology, Institute of Molecular Medicine, John Radcliffe Hospital, Oxford OX3 9DU, United Kingdom. e-mail: avincent@hammer.imm.ox.ac.uk

1. The absence of a fully developed lymphatic drainage;
2. The presence of an unusually tight endothelium within the inner walls of blood vessels in the brain and forming the blood-brain barrier (BBB), which normally expresses low levels of adhesion molecules and separates the nervous tissue from cells and soluble molecules present in the circulation;
3. The low expression of major histocompatibility complex (MHC) antigens in CNS tissue where MHC expression is restricted to perivascular and meningeal macrophages;
4. The fast clearance of infiltrating immune cells by apoptosis.

Conversely, the immune-privileged status of the CNS is not absolute; autoimmune reactions can occur within the nervous system because:

1. Alternative routes may allow diffusion of nervous system interstitial fluid into lymph and blood circulation [3-7]. This means that antigen-specific responses to CNS antigens can occur in the deep cervical lymph nodes after intracerebral antigen deposition, or in the spleen when antigens are deposited in the lateral ventricles that drain preferentially to the blood. The cerebral interstitial fluid is drained to the deep cervical lymph nodes via the cerebral perivascular spaces, and from the subarachnoid space of the olfactory lobes into the nasal submucosa. The cerebrospinal fluid (CSF) is also connected to the lymph nodes via prolongation of the subarachnoid space around the olfactory nerves leading into the interstitial spaces of the nasal submucosa.
2. The BBB is tight but does not form a barrier to activated T cells, irrespective of their antigen specificity [8-12]. In fact, activated T cells cross the BBB in order to perform a continuous immune surveillance of the nervous system. Granulocytes, monocytes and B cells, by contrast, can only cross the BBB through disrupted interendothelial tight junctions. However, soluble immunomodulatory molecules, such as antibodies, are not entirely excluded since small amounts do appear in the CSF, and soluble molecules can gain entry into the CNS parenchyma at specific regions such as the circumventricular organs of the hypothalamus.
3. MHC molecules can be expressed by neuronal cells if there is a strong proinflammatory stimulus [13]. During mild inflammation, MHC molecules (class I and II) are expressed solely by microglia, but during more marked inflammation neuroectodermal cells, such as astrocytes and ependymal cells, express class I and II MHC molecules. In addition, it has recently been shown that block of electrical activity in neurons induces MHC expression [14]. Moreover, during an autoimmune reaction, local production of chemokines can cause upregulation of MHC molecules and/or adhesion molecules on endothelial cells, thus regulating the infiltration of the nervous system by blood-borne immune cells.
4. Immune cells are cleared very rapidly from the nervous system [15-18]. Thirty percent to 40% of infiltrating T cells die by apoptosis, possibly through the Fas-Fas ligand pathway, or the tumour necrosis factor (TNF) receptor 1 functional signalling pathway, which is usually induced by high-dose soluble antigen. The fast clearance of immune cells from the nervous system is demonstrated by the

fact that within 24 h, more than three times the total number of infiltrating T cells can be cleared by apoptosis. Thus a continuous flux of T cells into the brain is necessary to keep an inflammatory process ongoing. The mechanisms of clearance of B cells and macrophages are still unknown.

Regulatory and Effector Mechanisms in Immune-Mediated Neurological Diseases

The pathological hallmark of immune-mediated disorders of the nervous system is the presence, within the target organ, of inflammatory infiltrates causing damage to the target cells and leading to functional impairment [19]. Inflammatory infiltrates usually contain autoreactive T and B cells, and pathogenic, non-antigen-specific, mononuclear cells. It is currently believed that antigen-reactive T cells provide the organ specificity of the pathogenic process and regulate the recirculation within the nervous system of the non-antigen-specific lymphocytes and monocytes, which act as effector cells by releasing toxic substances [20, 21]. T cells that are specific for neuronal components and mainly display the α/β T cell receptor (TCR), constitute the majority of the neuronal antigen-specific T cell population. Blood-borne activated macrophages producing neurotoxic substances (nitric oxide derivatives, pro-inflammatory cytokines, matrix metalloproteinases), B cells that produce antibodies against neuronal components, and γ/δ T cells represent the effector cell populations. However, the two different cell populations display overlapping functions: a minor proportion of α/β T cells that are specific for neuronal antigens show cytotoxic properties, whereas γ/δ T cells can contribute to effector cell recruitment (mainly macrophages) via pro-inflammatory cytokine or chemokine production. To further complicate the T-cell-mediated pathogenic scenario, it has recently been reported that both regulatory and effector cells can be cross-regulated by different subsets of T cells, including anti-TCR T cells and T cells that carry natural killer receptors.

General Principles for Measuring Neuronal Autoreactivity

The presence of circulating autoreactive T and/or B cells (or autoantibodies) against neuronal antigen is a prerequisite for determining their pathogenic role [22]. However, apparently autoreactive T and B cells are not only found in patients suffering from autoimmune neurological diseases, and such cells may be identified in healthy subjects [23, 24]. Thus, additional features have to be taken into consideration when the pathogenic role of an autoimmune reaction is suspected in neurology. With respect to T-cell-mediated diseases these features are: (1) association between certain MHC molecules and the neurological disease; (2) the association between the presence of neuronal T cell autoimmunity and an active phase of the disease; (3) improvement after immunosuppressive therapies; (4) the ability to reproduce at least part of the spectrum of the neurological disease in experimental animals by active immunisation or by systemic injection of appropriate antigen-specific T cells. In addition, with respect to antibody-mediated disorders, one should add (5) the therapeutic response to plasmapheresis or immunosuppressive treatments; (6) maternal-to-fetal transmission of transient neurological symptoms; and (7) ability to reproduce the neurological disease in experimental animals by injection, intravenously or intraperitoneally, of large amounts of the serum samples. In the latter – so-called "passive transfer" studies – an observational period of several days may be needed before full signs of disease develop in the animal. Even so, because of interspecies differences in antigens, and in possible accessibility problems, the disease may be subclinical in the experimental animal, and physiological and other tests may be needed to demonstrate functional effects [25].

Here, we discuss the ways of measuring autoimmunity to neuronal antigens in patients suffering from "putative" autoimmune neurological diseases. The study of the autoimmune T cell repertoire is less informative, from a clinical point of view, than the detection of autoantibodies because it can be easily performed only when the putative autoantigen is known. In this volume, since most of the diseases described have a strong B cell immune component, we will concentrate on assays for antibodies to neuronal antigens.

Autoantibodies in Neurological Diseases

Since a number of autoantigens that might induce autoimmune neurological disorders have been discovered in the past years, autoantibody assays have been used to diagnose, predict and explain autoimmune reactions [26-30]. Autoantibodies are divided between organ-specific and ubiquitous autoantibodies and several mechanisms have been proposed for their pathogenesis. There is only a partial correlation between the presence of autoantibodies in humans or animals and the risk of developing an autoimmune disease. Responses in animals vary according to genetic influence and extrapolation to humans is difficult. Therefore, in practice, autoantibody assays are not always predictive of autoimmune diseases, but can reflect autoimmune-mediated damage. Below we describe the techniques and assess the usefulness of different antibody assays, and comment on their significance.

Detection of Neuronal-Specific Antibodies

There are several ways to measure autoantibodies to neuronal antigens. They differ in terms of sensitivity and specificity, and the choice of the appropriate test has to be carefully made depending on the circumstances (for review see [31-38]).

Where the antigen is not known, but tissue specificity is suspected, immunohistochemistry is the main starting point, and can be followed by western blotting to try to identify relevant antigens [31]. In those cases in which there is a known or "candidate" antigen, the easiest techniques to use are immunoprecipitations of ^{125}I- or ^{35}S-labelled antigens [32] or enzyme-linked immunosorbent assay (ELISA)s [33, 34]. In these cases, where the pure antigen is available, one might also want to quantify antibody-forming B cells (AFC) by detection in situ [31].

Immunohistochemistry

In situ analysis of tissue sections by immunohistochemical methods is relatively simple, although technical aspects concerning freezing, sectioning, storing, fixing, incubating, and staining the tissues need to be considered before the results are interpreted. Classically, primary antibodies (i.e. patient's serum and/or CSF) are allowed to bind to conformational determinants (epitopes) of the antigen(s) present in tissue sections. Primary antibodies can be directly labelled with reporter molecules (e.g. fluorochromes, colloids, enzymes) to allow their detection, but in routine screening it is more usual to apply a second antibody conjugated to a reporter molecule. Alternatively, to increase sensitivity, one can use enzyme-antienzyme immune complexes, which achieve an amplification step (two-step procedure). Primary or secondary antibodies can also be derivatized with haptens (e.g. biotin, trinitrophenyl, digoxygenin), which can be specifically detected with their counterstructures (e.g. avidin or antibodies), resulting in more complex and usually more effective two- or three-step procedures. Immunohistochemistry can be performed with commercial kits, but to confirm specificity of the antibodies for the antigen it is sometimes necessary to use custom-made conjugates.

Immunohistochemistry is particularly useful as a first screen for antibodies, and is a sensitive technique to pick up most antibodies directed against intracellular components. Care must be taken, however, if the target is soluble in the cytoplasm or contained within intracellular particles. Permeabilisation of tissue sections with detergents, with or without paraformaldehyde fixation, can be used to ensure that particulate components, such as peptides or transmitters, are accessible but do not wash away during the procedure. As a general rule, immunohistochemistry is of limited use for detecting antibodies to membrane antigens, therefore it is important to remember that the antibodies that they detect may not be pathogenic in vivo.

Western Blotting

Western blotting on recombinant proteins and tissue extracts has often been used to detect antibodies, but it is less sensitive than radioimmunoprecipitation assays (RIA). After the separation by molecular weight and charge on a gel, the recombinant protein or tissue extract proteins are transferred onto a membrane (i.e. nitrocellulose or poly-vinylchloride) which is then incubated with the serum or other biological fluid to be tested. The antigen-antibody complex is then revealed using enzymatic or radiaoctive reactions. A limitation of these tests, leading to false negative results, is that they are usually used on tissues or protein extracts pretreated with reducing agents (particularly ionic detergents such as SDS, and often also with β-mercaptoethanol). These pretreatments are specifically intended to denature the proteins so that the individual protein subunits can be separated. Thus there is often loss of the conformational epitopes on a protein which might be crucial for antibody binding. However, this technique is useful when an antibody against an unknown auto-antigen is searched for. Furthermore, some technical modifications, such as reducing the detergents and running the proteins under "non-denaturing" conditions, can be used to avoid conformational changes in the target antigens. Some antibodies are very easy to detect with western blotting. It appears that, in general, those antibodies that are directed against intracellular antigens are often detected by immunohistochemistry and western blotting. Those antibodies that are against functional membrane proteins are usually much more difficult to detect by these methods. A similar method to western blotting is thin-layer chromatography followed by blotting [35]. This method is used for separating neuronal glycolipids (gangliosides) and for detection of antibodies binding to them.

Enzime-Linked Immunosorbent Assay

ELISA was first described by Engvall and Perlmann (1971) [33] and Schuurs and Van Weemen (1977) [34] and provides a safe and simple method of measuring antigen-specific and total immunoglobulins (Ig). ELISA falls under the category of heterogeneous immunosorbent assays in which (1) the bound and free fractions of Ig (ligand) are physically separated by a washing procedure and (2) the antigen to be detected or determined is either directly or indirectly physically attached to a solid phase. Various types of ELISA have been developed and described in the literature. The commonly used ones can be divided into different categories depending upon the following criteria:

1. Whether they are designed to measure, qualitatively or quantitatively, an antigen or an antibody;
2. Whether they involve direct binding of labelled antibody or more complex procedures such as sandwich ELISAs, competition between antibody and labelled antibody, or more complicated amplifications;
3. The nature of the detection substrate employed (fluorogenic, chromogenic, luminogenic) [36];
4. The amounts of reactants used [macro-ELISA performed in a tube, micro-ELISA employing a microtitre plate, or micro-ELISA needing micro quantities (5 µl) of sample performed in Terasaki trays, the Terasaki-ELISA];
5. The nature of the solid phase used to adsorb the coating protein (glass, polypropylene, PVC, polystyrene, polycarbonate, nitrocellulose, silicone, etc.).

The majority of ELISAs are performed in polystyrene microtitre plates using commercially available equipment, facilitating easy and automatic handling of the system. They can be used to determine the concentration of immunoglobulins (IgM, IgA, IgG) in moderately sensitive ELISAs. However, for simultaneous detection of a number of Ig classes (especially rare ones like IgD and IgE), autoantibodies or anti-idiotypic antibodies, a fairly large serum sample is needed.

ELISAs using recombinant proteins or purified gangliosides (glycolipids) can be very useful for measuring antibodies since they are easily handled and can be performed almost everywhere. One major disadvantage of these tests, however, is that the antigen must be available in soluble form, and that the sensitivity of the systems is such that they seldom detect antibodies (or antigens) in the picogram

(10-12 pg) range. A recent use of ELISA employing recombinant extracellular domains of the protein attached to microtitre plates is in measuring antibodies to the muscle-specific kinase, MuSK [37].

Radioimmunoprecipitation Assays of Radio-Labelled Proteins

RIA is a very sensitive technique that has been employed in a number of ways [32]. The technique relies on radio-labelling of the antigen. This can be achieved by direct labelling of recombinant proteins by ^{125}I, or of incorporation of ^{35}S-methionine into protein made in situ by cell-free synthesis. Alternatively, the labelling can be achieved by binding of radioactive ligands that have high affinity for the antigen, such as ^{125}I-α-bungarotoxin for radio-labelling acetylcholine receptor (AChR). However, one needs to consider that (1) the radioisotopes or radio-ligand may interfere with binding of the test antibody, leading to false negative results; (2) the putative antigen cannot be labelled directly if it lacks residues (usually tyrosines) suitable for radio-labelling; and (3) serum autoantibodies or heterophilic antibodies might interfere with the radioassay. Nevertheless, this is, so far, the simplest and most sensitive approach for detecting antibodies to AChR, voltage-gated calcium channels (VGCCs), voltage-gated potassium channels (VGKCs) (see [38]), glutamic acid decarboxylase (GAD) [28], and more recently botulinum toxin and interferon (IFN)β.

Functional Assays

Sometimes functional assays on cell lines and/or tissue slices are necessary to measure the presence of autoantibodies and to begin to deduce their targets. However, these tests require an appropriate cell line to test, and not all antibodies interfere with function even when their antigen is present on the cell surface. Nevertheless, this approach has been most important for showing that antibodies in the Lambert-Eaton myasthenic syndrome (LEMS) were capable of interfering with VGCC function on cultured neuronal and small-cell lung cancer cell lines [39]; for demonstrating the mechanisms of antibody-induced loss of AChRs in myasthenia gravis (MG) sera [40]; and for showing that sera from MG patients without AChR antibodies are able to alter AChR function in muscle cell lines, thus confirming that they do contain antibodies binding to muscle antigens [41].

Binding to Cell Lines and Use of Fluorescence-Activated Cell Sorting

One way to detect antibodies to tissue-specific surface antigens or to known antigens is to use cell lines expressing the antigen and apply serum and fluorescently labelled second antibodies. This has been used, for instance, to look for antibodies to glutamate receptors in Rasmussen's encephalitis [42], and for antibodies to muscle cell antigens in AChR-antibody-negative MG [43]. The fluorescence-activated cell sorting (FACS) machine offers an objective way of measuring surface fluorescence which can be quantified and evaluated independently.

Further Analysis of Antibodies

Specific Antibody-Forming B Cells

A direct immunoenzyme approach for the detection of antigen-specific antibody-forming B cells (AFC) in situ can be used [31]. In animal studies with model antigens, and in human studies with viral antigens and autoantigens, specificity of B cells in tissue sections can be demonstrated after incubation with antigen-enzyme conjugates, and their isotype determined using an anti-Ig-(Fc-chain-specific)-enzyme conjugate followed by appropriate enzyme chemistry. To identify cells in tissue sections, or cultured in vitro, that are specifically secreting antibodies, one can use hapten-enzyme and hapten-carrier conjugates to identify B cells secreting antibodies against specific antigens. The main strength in the method for detecting specific AFC lies in the fact that only those cells actively involved in the immune response against the defined antigen or peptide epitope are identified.

Peptide Libraries

The analysis of antibody responses to auto- and xenoantigens has been pursued by immunologists interested in a diverse array of scientific questions related to the mechanisms of antibody repertoire diversification, the protective and pathogenic actions of antibodies, the molecular changes in the genes encoding antibodies and the dissection of the epitopes recognised by cells of the immune system. Random peptide libraries present short peptide sequences on the surface of bacterial plasmids [44]. The detection of the peptides that bind to a particular antibody can be made by growing the bacteria on agar plates, immunoblotting the plates with nitrocellulose, and then detecting the position of the synthesised antigenic peptides with the specific antibody. The bacteria expressing this peptide can then be cloned further and sequenced to identify the target. Alternatively, the antibody can be immobilised on protein A beads and these used to select the peptide-expressing plasmids that bind to the antibody. In both cases, it is obviously preferable to have a purified antibody.

Production of Cloned Antibodies

In the last two decades, immortalisation of the antibody-producing cells by somatic hybridization or viral transformation has frequently been the method of choice to obtain monoclonal cell lines secreting a single species of antibody [45]. Although extremely useful, this approach permits only a limited sampling of the immune repertoire and has posed severe technical limitations in organisms other than the mouse. In order to identify the antibodies made, specific assays must be applied. Some success has recently been achieved with cloning and selection of antibodies binding specifically to human AChR. This has been performed by making a RNA library from a tissue source that contains the AFC, e.g. the thymus in MG, subjecting the heavy and light chains from the B cell antibody repertoire to polymerase chain reaction and then randomly combining these in tandem and expressing them in the peptide expression plasmids. The bacteria producing specific recombined single chain Fab antibodies can then be selected with radio-labelled antigen or solid phase antigen (see peptide libraries, above) [46, 47]. Using this approach, antibodies specific for fetal human AChR have been cloned from the thymus of two women whose sera contained predominantly fetal AChR antibodies [48].

Evaluating the Pathogenic Role of Antibodies

As indicated in the introduction to this chapter, there are various clues that demonstrate a role for antibodies in autoimmune diseases. The most effective way of confirming an antibody-mediated pathology is to perform plasma exchange. This was clearly shown to improve patients with MG, LEMS, acquired neuromyotonia and some cases of peripheral inflammatory neuropathies. Moreover, importantly, it is sometimes also effective in CNS disorders with known antibodies, such as Rasmussen's encephalitis and Morvan's syndrome, suggesting that it should not be neglected in presumed autoimmune CNS diseases (Tables 1 and 2, and see relevant chapters).

Passive transfer of immunoglobulins to mice (or occasionally other species) is also a most important approach to demonstrate the presence of autoantibodies. Injection schedules are usually up to 1 ml daily of plasma or purified IgG per mouse intraperitoneally [49]. Using purified and concentrated IgG (around 40 mg/ml), it is possible to attain levels of human IgG in the mouse plasma that are similar to those in the donor [50]. However, as mentioned above, the physiological effects may not be so evident as in the patient, probably because of reduced cross-reactivity with mouse proteins, higher turnover of the foreign antibodies, inefficient involvement of mouse complement, and resistance of the mouse nervous system to manifestations of the disease. Therefore, subtle tests of nerve or muscle function and in vitro analysis should be performed.

The use of passive transfer in CNS disorders is still not clear. In general there have been few attempts to achieve CNS abnormalities after systemic injection of antibodies, and relatively few reports of intrathecal injection. Although the latter approach clearly gets round the problem of the BBB to some extent, it does not reproduce the nature of the naturally occurring disease. Thus new approaches, and more careful analysis, are required to establish models of antibody-mediated CNS disorders.

In order to bring together the evidence discussed above, and to demonstrate the usefulness of the techniques described, we have put together two tables representing most of the conditions included in this volume (Tables 1, 2). These tables should help to illustrate some of the important and helpful principles that have derived from the study of antibody-mediated disorders.

It is clear from differences between Table 1 and Table 2, and in the chapters that

Table 1. Clinical and immunological characteristics of well-characterised autoantibody-mediated neurological disorders of the peripheral nervous system

Disease	Pathology	Antigen	Diagnosis	Treatment
Myasthenia gravis	Loss of AChRs	AChRs	Anti-AChR	Responds to immunosuppression and thymectomy
	Loss of MuSK? Loss of AChRs?	MuSK	Anti-MuSK	Responds to immunosuppression
Lambert-Eaton syndrome	Loss of VGCCs	VGCC P/Q-type	Anti-VGCC	Plasma exchange, immunosuppression and intravenous immunoglobulin
Acquired neuromyotonia	Loss of VGKCs	VGKC on motor nerve	Anti-VGKC	Responds to anti-epileptic drugs and immunosuppression
Guillaine-Barré syndrome	Demyelination and/or axonal damage	GM1, GD1b and other glycolipids	Anti-GM1 in <50%	Plasma exchange and intravenous immunoglobulin
Miller-Fisher syndrome	Demyelination of ocular motor nerves?	GQ1b and other polysialylated glycolipids	Anti-GQ1b in >90%	Plasma exchange and intravenous immunoglobulin

AChR, acetylcholine receptor; *VGCC*, voltage-gated calcium channel; *VGKC*, voltage-gated potassium channel

Table 2. Clinical and immunological characteristics of possible autoantibody-mediated neurological disorders of the central nervous system

Disease	Pathology	Antigen	Diagnosis	Treatment
Rasmussen's encephalitis	Inflammation of one cerebral hemisphere	GluR3	Clinical	Immunosuppression may be helpful
Limbic encephalitis	Inflammation of hippocampus	Hu, Ma2	Anti-Hu antibodies	Not usually effective
		VGKC	Anti-VGKC antibodies	Plasma exchange
Cerebellar ataxia	Loss of Purkinje cells	Yo	Anti-Yo antibodies	Not usually helpful
	Not known	VGCC	Anti-VGCC antibodies	Does plasma exchange prevent progression?
	Not known	GAD	Anti-GAD antibodies	Immunotherapies might be helpful
Stiff-man syndrome	Loss of inhibitory neurones?	GAD	Anti-GAD antibodies	Immunotherapies might be helpful
Opsoclonus-myoclonus	Not known but could affect brain stem neurones	Not known	Not known	Definitely helpful but needed throughout childhood
Bieckerstaff's encephalitis	Reversible inflammation in brain stem	GQ1b	Anti-GQ1b antibodies and imaging	Not clear yet

VGKC, Voltage-gated potassium channel; *VGCC*, voltage-gated calcium channel; *GAD*, glutamic acid decarboxylase follow, that antibodies to peripheral targets are more closely related to the neurological conditions, and that immunotherapies are effective in treating these patients. Whether the same will turn out to be the case in those CNS disorders that are associated with specific antibodies to cell surface antigens, such as VGCCs and VGKCs, remains to be seen, although preliminary evidence suggests that this will be the case.

Conclusions

The presence of circulating autoantibodies (and also autoreactive T or B cells) against neuronal antigens is not exclusive to patients suffering from autoimmune disease of the nervous system. However, there are now many useful ways of detecting antibodies of high affinity and specificity that are likely to be directly involved in causing disease, or that are very helpful markers for paraneoplastic conditions as will be clear in the chapters that follow. Further work is now needed to begin to apply these principles to disorders of the CNS, thus defining diseases that might be amenable to immunological treatments.

References

1. Barker CF, Billingham RE (1977) Immunologically privileged sites. Adv Immunol 25:1-54
2. Cousins SW, McCabe MM, Danielpour D, Streilein JW (1991) Identification of transforming growth factor-beta as an immunosuppressive factor in aqueous humor. Invest Ophthalmol Vis Sci 32:2201-2211
3. Bradbury MW, Cole DF (1980) The role of the lymphatic system in drainage of cerebrospinal fluid and aqueous humour. J Physiol 299:353-365
4. Widner H, Moller G, Johansson BB (1988) Immune response in deep cervical lymph nodes and spleen in the mouse after antigen deposition in different intracerebral sites. Scand J Immunol 28:563-571
5. Cserr HF, Harling-Berg CJ, Knopf PM (1992) Drainage of brain extracellular fluid into blood and deep cervical lymph and its immunological significance. Brain Pathol 2:269-276
6. Bradbury MW, Cserr HF, Westrop RJ (1981) Drainage of cerebral interstitial fluid into deep cervical lymph of the rabbit. Am J Physiol 240:F329-336
7. Bradbury MW, Westrop RJ (1983) Factors influencing exit of substances from cerebrospinal fluid into deep cervical lymph of the rabbit. J Physiol 339:519-534
8. Hickey WF (1991) Migration of hematogenous cells through the blood-brain barrier and the initiation of CNS inflammation. Brain Pathol 1:97-105
9. Wekerle H, Engelhardt B, Risau W, Meyermann R (1991) Interaction of T lymphocytes with cerebral endothelial cells in vitro. Brain Pathol 1:107-114
10. Raine CS, Cannella B, Duijvestijn AM, Cross AH (1990) Homing to central nervous system vasculature by antigen-specific lymphocytes. II. Lymphocyte/endothelial cell adhesion during the initial stages of autoimmune demyelination. Lab Invest 63:476-489
11. Wisniewski HM, Lossinsky AS (1991) Structural and functional aspects of the interaction of inflammatory cells with the blood-brain barrier in experimental brain inflammation. Brain Pathol 1:89-96

12. Cross AH, Raine CS (1991) Central nervous system endothelial cell-polymorphonuclear cell interactions during autoimmune demyelination. Am J Pathol 139:1401-1409
13. Vass K, Lassmann H (1990) Intrathecal application of interferon gamma. Progressive appearance of MHC antigens within the rat nervous system. Am J Pathol 137:789-800
14. Neumann H, Cavalie A, Jenne DE, Wekerle H (1995) Induction of MHC class I genes in neurons. Science 269:549-552
15. Pender MP, Nguyen KB, McCombe PA, Kerr JF (1991) Apoptosis in the nervous system in experimental allergic encephalomyelitis. J Neurol Sci 104:81-87
16. Schmied M, Breitschopf H, Gold R et al (1993) Apoptosis of T lymphocytes in experimental autoimmune encephalomyelitis. Evidence for programmed cell death as a mechanism to control inflammation in the brain. Am J Pathol 143:446-452
17. Griffith TS, Ferguson TA (1997) The role of FasL-induced apoptosis in immune privilege. Immunol Today 18:240-244
18. Weishaupt A, Gold R, Gaupp S et al (1997) Antigen therapy eliminates T cell inflammation by apoptosis: effective treatment of experimental autoimmune neuritis with recombinant myelin protein P2. Proc Natl Acad Sci USA 94:1338-1343
19. Aloisi F, Ria F, Adorini L (2000) Regulation of T cell responses by central nervous APC: different roles for microglia and astrocytes. Immunol Today 21:141-147
20. Martino G, Hartung HP (1999) Immunopathogenesis of multiple sclerosis: the role of T cells. Curr Opin Neurol 12:309-321
21. Kieseier BC, Storch MK, Archelos JJ et al (1999) Effector pathways in immune mediated central nervous system demyelination. Curr Opin Neurol 12:323-336
22. Parry SL, Hall FC, Olson J et al (1998) Bctivity versus autoaggression: a different perspective on human autoantigens. Curr Opin Immunol 10:663-668
23. Zinkernagel RM (2000) What is missing in immunology to understand immunity? Nat Immunol 1:181-185
24. Vincent A, Lily O, Palace J (1999) Pathogenic autoantibodies to neuronal proteins in neurological disorders. J Neuroimmunol 100:169-180
25. Vincent A (1999) Antibodies to ion channels in paraneoplastic disorders. Brain Pathol 9:285-291
26. Archelos JJ, Hartung HP (2000) Pathogenetic role of autoantibodies in neurological diseases. Trends Neurosci 23:317–327
27. Giometto B, Tavolato B, Graus F (1999) Autoimmunity in paraneoplastic neurological syndromes. Brain Pathol 9:261-273
28. Lang B, Vincent A (1999) Autoimmunity to ion-channels and other proteins in paraneoplastic disorders. Curr Opin Immunol 8:865-871
29. Vincent A, Honnorat J, Antoine JC et al (1998) Autoimmunity in paraneoplastic neurological disorders. J Neuroimmunol 84:105-109
30. Moll JW, Antoine JC, Brashear HR et al (1995) Guidelines on the detection of paraneoplastic anti-neuronal-specific antibodies: report from the Workshop to the Fourth Meeting of the International Society of Neuro-Immunology on Paraneoplastic Neurological Disease, 22-23 October 1994, Rotterdam, The Netherlands. Neurology 45:1937-1941
31. Lefkovits I (ed) (1996) Immunology methods manual: The comprehensive sourcebook of techniques. Academic Press, San Diego
32. Berson SA, Yalow RS (1968) General principles of radioimmunoassay. Clin Chim Acta 22:51-69
33. Engvall E, Perlman P (1971) Enzyme-linked immunosorbent assay (ELISA). Quantitative assay of immunoglobulin G. Immunochemistry 8:871-874
34. Schuurs AH, Van Weemen BK (1977) Enzyme-immunoassay. Clin Chim Acta 8:1-40

35. Pataki G (1967) Thin-layer chromatography of amino acids. Chromatogr Rev 9:23-36
36. Roda A, Pasini P, Guardigli M et al (2000) Bio- and chemiluminescence in bioanalysis. Fresenius J Anal Chem 366:752-759
37. Hoch W, McConville J, Helms S et al (2001) Autoantibodies to the receptor tyrosine kinase MuSK in patients with myasthenia gravis without acetylcholine receptor antibodies. Nat Med 7: 365-368
38. Vincent A, Beeson D, Lang B (2000) Molecular targets for autoimmune and genetic disorders of neuromuscular transmission. Eur J Biochem. 267:6717-6728
39. Roberts A, Perera S, Lang B et al (1985) Paraneoplastic myasthenic syndrome IgG inhibits 45Ca2+ flux in a human small cell carcinoma line. Nature 317:737-739
40. Drachman DB, Adams RN, Josifek LF, Self SG (1982) Functional activities of autoantibodies to acetylcholine receptors and the clinical severity of myasthenia gravis. N Engl J Med 307:769-775
41. Yamamoto T, Vincent A, Ciulla TA et al (1991) Seronegative myasthenia gravis: a plasma factor inhibiting agonist-induced acetylcholine receptor function copurifies with IgM. Ann Neurol 30: 550-557
42. Rogers SW, Andrews PI, Gahring LC et al (1994) Autoantibodies to glutamate receptor GluR3 in Rasmussen's encephalitis. Science 265:648-651
43. Blaes F, Beeson D, Plested P et al (2000) IgG from "seronegative" myasthenia gravis patients binds to a muscle cell line, TE671, but not to human acetylcholine receptor. Ann Neurol 47:504-510
44. Li M (2000) Applications of display technology in protein analysis. Nat Biotechnol 18:1251-1256
45. Larrick JW, Fry KE (1991) Recombinant antibodies. Hum Antibodies Hybridomas 2:172-189
46. Graus YF, de Baets MH, Parren PW et al (1997) Human anti-nicotinic acetylcholine receptor recombinant Fab fragments isolated from thymus-derived phage display libraries from myasthenia gravis patients reflect predominant specificities in serum and block the action of pathogenic serum antibodies. J Immunol 158:1919-1929
47. Farrar J, Portolano S, Willcox N et al (1997) Diverse Fab specific for acetylcholine receptor epitopes from a myasthenia gravis thymus combinatorial library. Int Immunol 9:1311-1318
48. Matthews I, Farrar J, McLachlan S et al (1998) Production of Fab fragments against the human acetylcholine receptor from myasthenia gravis thymus lambda and kappa phage libraries. Ann N Y Acad Sci 841:418-421
49. Toyka KV, Drachman DB, Griffin DE et al (1977) Myasthenia gravis: study of humoral immune mechanisms by passive transfer to mice. New Eng J Med 296:125-131
50. Mossman S, Vincent A, Newsom-Davis J (1988) Passive transfer of myasthenia gravis by immunoglobulins: lack of correlation between antibody bound, acetylcholine receptor loss and transmission defect. J Neurol Sci 84:15-28

Chapter 1

New Antibodies to Neuronal and Muscle Antigens

A. VINCENT[1], C. BUCKLEY[1], P. DALTON[1], L. CLOVER[1], R. LIGUORI[2], P. MONTAGNA[2], J. MCCONVILLE[1], W. HOCH[3]

The role of antibodies to ion channels at the neuromuscular junction is now well established. Antibodies binding to the muscle isoform of the acetylcholine receptor (AChR) are present in the majority of patients with myasthenia gravis (MG; see Chaps. 2, 4), and antibodies binding to the voltage-gated calcium channel in the Lambert-Eaton myasthenic syndrome (LEMS; see Chap. 3). However, the recognition of the role these antibodies play in the pathogenesis of the disorders, and their usefulness for their diagnosis, stimulated the search for other antibody-mediated diseases of the peripheral nervous system, and also raised questions concerning their possible involvement in central nervous system (CNS) diseases (see the Introduction to this volume).

Here we will briefly summarise new findings that show (1) that antibodies to a developmental protein, MuSK, are present in some cases of MG without AChR antibodies; (2) that antibodies to fetal AChR and probably other muscle or neuronal antigens can affect fetal development; and (3) that antibodies to voltage-gated calcium and potassium channels can sometimes be associated with CNS disease.

Seronegative Myasthenia Gravis

About 10%-20% of MG patients, termed seronegative MG (SNMG) patients, do not have detectable AChR antibodies by the conventional immunoprecipitation assay (see volume Introduction). Although some of these patients' sera may be positive for antibodies that directly inhibit AChR function, or for "modulating" antibodies that increase the degradation rate of AChRs on cell lines, the majority do not have these reactivities [1] and, in fact, do not appear to bind to AChRs at all [2]. This has recently been confirmed by a study in which binding of IgG antibodies to a muscle-like cell line, TE671, which expresses AChRs, was quantitative-

[1] Neurosciences Group, Department of Clinical Neurology, Institute of Molecular Medicine, John Radcliffe Hospital, Oxford OX3 9DU, United Kingdom.
e-mail: avincent@hammer.imm.ox.ac.uk
[2] Institute of Neurology, University of Bologna, Italy
[3] Max Planck Institute for Development Biology, Tübingen, Germany

ly measured by fluorescence-activated cell sorting. SNMG IgG antibodies bound strongly to the TE671 cells, but they did not bind to human embryonic kidney cells that had been engineered to express human AChRs [3], and must therefore be binding to another membrane protein. A previous result on a mouse cell line could be interpreted similarly [4].

Other observations had suggested that when the antibodies in SNMG sera bound to the TE671 cells, they triggered a second messenger pathway that resulted in AChR phosphorylation. This suggestion was based on the fact that application of the sera inhibited AChR function [5], but this inhibition was transient and associated with increased incorporation of ^{32}P into AChRs [6]. Electrophysiological studies were also consistent with the idea that the SNMG antibodies bound to a distinct cell surface protein that could indirectly lead to inhibition of AChR function. TE671 cells were patch-clamped in the "whole cell" mode and AChRs' openings recorded within the patch pipette. When the serum was applied to the cell, outside the patched area, the opening frequencies of the AChRs within the patch were reduced (by two of three sera tested [6]).

Collectively, these observations suggested that the target for the SNMG antibodies was a cell surface membrane protein that could activate second messengers. Since serine and threonine kinase inhibitors did not influence the action of SNMG sera [7], a tyrosine kinase appeared to be likely. A possible candidate was the tyrosine kinase receptor MuSK (muscle specific kinase). MuSK is an essential component of the neuromuscular junction during development, when it orchestrates AChR clustering (reviewed in [8]). Agrin released from the motor nerve binds to MuSK, or via an associated protein that has not yet been identified, and this causes autophosphorylation of MuSK, phosphorylation of the receptor-aggregating protein RAPsyn, phosphorylation of the AChR and AChR clustering. The clustering activity of agrin can be observed in myotube cultures in vitro (e.g. [9]).

IgG antibodies from SNMG sera bind to the surface of COS7 cells expressing rat MuSK, bind to the extracellular domain of rat MuSK in enzyme-linked immunosorbent assays, and inhibit the agrin-induced clustering of mouse AChRs in C2C12 myotubes [10]. These MuSK antibodies are present in about 70% of SNMG patients tested so far, and are not present in MG patients with AChR antibodies. It is too early to say whether the sensitivity of the assay will improve when human MuSK is used, and how many patients with SNMG will be positive in diagnostic assays. However, positive values have been found in some AChR antibody-negative sera sent for routine analysis (J. McConville, W. Hoch, N. Lawrence, A. Vincent, unpublished data), and it is hoped that measurement of MuSK antibodies will provide a useful confirmatory test for the diagnosis of MG in these patients.

Antibodies and Development

Developmental abnormalities are relatively common, some degree of congenital defect being found in up to 5% of babies. Although many well-characterised conditions are known to be genetically determined, many others have no known cause. A typical syndrome of developmental abnormalities is arthrogryposis multiplex congenita (AMC). This describes the presence of multiple joint contractures in babies at birth, or noted in utero. It is often associated with other deformities such as small chin (micrognathia) and hypoplastic lungs, in which case the condition can be fatal. There are many known genetic disorders associated with arthrogryposis, but the majority of cases are unexplained. The clinical aspects are reviewed in [11].

Arthrogryposis Multiplex Congenita

The underlying cause of AMC is thought to be lack of fetal movement in utero [12, 13], and it can therefore be caused by any situation leading to fetal paralysis or restriction of movement. It is probable that the lack of movement has to persist for several weeks for AMC to develop, and once the baby is born the condition should not progress if movements are initiated. In a small number of reported cases, AMC is caused by maternal antibodies to the AChR (see Chap. 2), sometimes in association with typical signs of neonatal MG in the baby, but often presenting as severe AMC with complete fetal paralysis and failure to breathe. Because of the marked hypoplasia of the lungs in these cases, the baby may not survive.

The antibodies associated with AMC are highly specific for the fetal isoform of the AChR. This is a feature also of other patients with MG, including those women who transmit neonatal MG to their babies [14]. The distinguishing feature in mothers of AMC babies is that the antibodies not only bind to sites on the fetal AChR but also potently inhibit its physiological function [15]. These inhibitory antibodies are much less common in MG patients in general, even those with neonatal MG (unpublished observations), and are thought to be responsible for the fetal paralysis which results in AMC.

A hallmark of AMC associated with maternal antibodies to the AChR is that the condition tends to recur with each pregnancy [16], unless the mother is treated or other preventive measures are undertaken. Thus, most of the reported cases had several consecutive affected babies. However, treatment of the mother for her MG, if present, may allow her to have an unaffected baby subsequently [17] (S.M. Huson, J. Newsom-Davis and A. Vincent, in preparation). Another striking feature of these cases is that the deformities noted at postmortem include a wide spectrum of abnormalities including cleft tongue and palate, abnormal genitalia, achalasia of the oesophagus, as well as spinal, cranial and brain defects [18]. Thus, either there are other antibodies in these women that are responsible for some of the varied, and not necessarily recurrent, defects, in other organs; or, alternatively, the lack of fetal movement caused by the AChR antibodies can lead indirectly to the other features. For instance, it is quite possible that lack of swallowing activity during gestation could lead not only to small jaw and other facial abnormalities, but also to defects in oesophageal patency causing oesophageal achalasia. Whether lack of muscle activity could, by itself, cause cerebral and cerebellar atrophy, as found in some cases, is less clear.

Another aspect of maternal-antibody-mediated AMC is that the mother may herself be asymptomatic [16], or the diagnosis of MG may not be made until after the birth of affected babies. Indeed, in many of the reported cases the mothers have had one apparently normal baby before several consecutive affected babies (e.g. [16]). This means that it is worth excluding maternal MG in any mother who has a baby with AMC. In fact, since these mothers have been studied in detail, a large number of sera have been obtained from apparently healthy women with AMC babies and tested for AChR antibodies. Although very few are positive, we have identified high levels of AChR antibodies, affecting fetal AChR function, in two apparently healthy mothers who have had one affected baby each. Of particular interest, in addition, is that another 25 sera out of approximately 200 tested are positive for antibodies binding to muscle or neuronal cell lines, and a few of these also inhibit fetal AChR function. One serum binds strongly to chondrocytes in developing bone (P. Dalton and A. Vincent, unpublished observations).

It is difficult to assess whether the antibody reactivity that has been found is pathogenically relevant, since in vitro tests tell one little about the action of the antibodies on intact tissue. In order to be able to test the potential pathogenicity of maternal serum antibodies, we established a passive transfer model of AMC, using a similar protocol to that used to demonstrate the effects of MG antibodies (see Introduction to this volume). Maternal plasma or IgG was injected daily into pregnant mice. The human AChR antibodies were transported surprisingly efficiently to the mouse fetuses resulting in levels of anti-AChR in the fetuses that were similar to those achieved in the pregnant mouse's own serum [19]. A high proportion of the mouse fetuses, born to mothers injected with antibodies from human mothers of AMC babies were born paralysed and with fixed joint deformities. These were surprisingly similar to those found in the human condition. Less marked deformities were found in babies treated in utero with antibodies from a woman who was healthy and AChR antibody-negative, but had had four consecutive babies affected by AMC [19]. This maternal-to-fetal passive transfer model can now be used to test the effect of serum antibodies from other mothers of AMC babies.

Origin of the Anti-Fetal Response

One question is whether the immune response to fetal antigens occurs in the thymus, as it appears to do in early-onset MG (see Chap. 4), or whether it occurs in secondary lymphoid tissues. Matthews et al. [20] made a combinatorial single chain Fab library from mRNA expressed in the thymus of two young women with MG and AMC babies. The Fabs were selected for AChR reactivity, sequenced and characterised. The Fabs were heterogenous in their sequences, but derived from a limited number of IgG heavy-chain genes which had undergone extensive somatic mutation, indicating a germ line origin and affinity maturation of the response. Strikingly, all but one of the Fabs from these two women bound to the fetal-specific AChR gamma subunit, and did not bind to the adult AChR. Disappointingly, none of them inhibited AChR function. However, the explanation probably lies in the fact that, despite the potency of these antibodies in maternal AMC sera, they do not form a substantial proportion of the total AChR antibody repertoire (A. Vincent, unpublished observations).

Pregnancy and Autoimmunity

These observations raise questions concerning the role of pregnancy in initiating maternal immunity towards the fetus. The role of maternal antibodies in Rhesus disease of the newborn is well known, and usually these antibodies do not affect the first-born. How often do women begin to react against fetal AChR during or after their first pregnancy? We studied this question by looking for antibodies inhibiting the function of the fetal AChR (similar to those in AMC mothers) in sera from women who developed MG either before pregnancy (non-parous) or during or after their first pregnancy (parous). There was a significantly larger proportion of antibodies inhibiting fetal AChR function in the parous women (Fig. 1); these antibodies were not as potent as in AMC mothers but were infrequent in non-parous women, and absent in men (A. Vincent, I. Matthews, N. Willcox, in preparation). This suggests that these inhibitory antibodies are in some way initiated by pregnancy, and that in these women with MG, the maternal

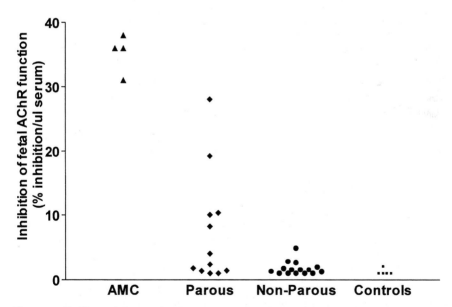

Fig. 1. Antibodies inhibiting the function of fetal acetylcholine receptors in different groups of myasthenia gravis patients and mothers of arthrogryposis babies (*AMC*). Antibodies that can inhibit the ion channel function of the receptor are more common in women who develop MG during or after pregnancy (*Parous*) than in those who develop it before pregnancy (*Non-Parous*)

disease might follow an anti-fetal response. Interestingly, there is growing evidence for the presence of cells of fetal origin in the mother, and it seems likely that these cells, perhaps in the context of a partially foreign major histocompatibility complex (MHC) background, can stimulate the maternal immune system. Nevertheless, the number of women who develop MG during or shortly after their first pregnancy is quite small, and most of the typical early-onset MG patients (onset below 40 years of age) develop MG in their teens. It will be interesting to look for fetal-specific antibodies in women who develop MG or other autoimmune diseases in later years, particularly as some conditions such as scleroderma are now thought to be associated with persistence of fetal cells in the maternal skin [21].

Other Developmental Disorders Might be Caused by Maternal Antibodies

The passive transfer model can now be used to test for pathogenic antibodies in serum from women with AMC babies who do not have AChR antibodies, and also to test other conditions in which maternal antibodies might be important. For instance, dyslexia and autism are both thought to be associated with higher familial incidence of autoimmunity, and dyslexia shows linkage to genes in the HLA regions. Although recurrent affected children with these conditions are relatively rare, some families have up to four affected children without any evidence to suggest a dominant or recessive inheritance. Thus we have begun to test the possibilty that these cases of developmental disorders might be due to the transfer of maternal antibodies to the developing fetus (A. Vincent, R. Deacon, P. Dalton et al, manuscript submitted for publication).

Conclusions

There are many possible routes to developmental disorders, including those conditions that present later in childhood or adolescence. Our understanding of the maternal immune response during pregnancy is just beginning (Chap. 2). It is now clear that not only do fetal cells reach the maternal circulation, but maternal cells, including possibly fetal cells from previous pregnancies, can reach the fetus. Since paternal and maternal antigens differ with respect to polymorphic proteins such as HLA, red cell antigens, platelet antigens etc, it would not be surprising if the mother becomes sensitised to fetal antigens in some cases. Understanding the target of these antibodies and how they cause fetal developmental abnormalities will require much painstaking work, and will rely heavily on work on development by developmental biologists. In this respect the chapter by Cossu (Chap. 5) demonstrates how many different pathways are involved in the development of muscle, and his review of the signalling pathways should suggest possible targets for maternal antibodies that could affect development of different muscle groups.

Disorders Associated with Antibodies to Voltage-Gated Calcium and Potassium Channels

VGCC Antibodies and Cerebellar Ataxia

Antibodies to voltage-gated calcium channels (VGCC) are present in about 90% of patients with LEMS (see Chap. 3) and are highly specific for the disease. The assay for VGCC antibodies depends on extraction of the ion channels from human or rabbit cerebellum, labelling with ^{125}I-conotoxin MVIIC, and immunoprecipitation with the patient's serum (see Introduction to this volume). The VGCCs identified by ^{125}I-conotoxin MVIIC are the so-called P/Q-type VGCCs that were defined by their presence in cerebellar Purkinje cells, and LEMS serum antibodies interfere with VGCC function on cerebellar Purkinje and granular cells in culture [22]. It would therefore not be surprising if some patients with LEMS had cerebellar symptoms. A small proportion of LEMS patients have signs of cerebellar involvement, and some patients who present with cerebellar ataxia turn out to have VGCC antibodies and LEMS, either with or without Hu antibodies (see Chap. 7). In a recent study of sera sent for investigation of paraneoplastic antibodies, three individuals with cerebellar ataxia were found to have VGCC antbodies without anti-Hu, and each of these patients was subsequently found to have a lung cancer [23].

Patients with VGCC antibodies and cerebellar symptoms often have detectable VGCC antibodies in their cerebrospinal fluid, although the amounts present are compatible with simple diffusion from the serum rather than evidence of intrathecal synthesis (F. Graus, B. Lang and A. Vincent, in preparation). Further studies are required to try to understand why some patients' VGCC antibodies cause CNS involvement and whether effective immunotherapies might be able to improve the ataxia, or at least halt its progress.

VGKC Antibodies, Neuromyotonia and Central Nervous System Disorders

Acquired neuromyotonia (NMT), also called Isaac's syndrome and many other names (see [24]), is defined as the presence of spontaneous or repetitive motor unit potentials leading to muscle fasciculations, cramps and pseudomyokymia. Another condition, the cramp fasciculation syndrome, may be a less severe version of the same conditions (Hart IK, Maddison P, Vincent A, manuscript submitted for publication). By definition, the bursts of motor unit potentials occuring at high intraburst frequencies, detected by electromyography of affected muscles, are generated in the distal motor nerve since they are abolished by curare, and not affected by proximal nerve block or sleep. Similar activity may occur in a variety of peripheral nerve disorders and are likely to be caused by different motor nerve pathologies [25]. Sensory problems are not usually a major feature of NMT, but are reported by about 15% of patients. Excessive sweating is a common and characteristic finding.

Pathophysiology

The condition is thought to be caused by loss of the VGKCs that are responsible for depolarising the motor nerve membrane potential after each action potential. Similar clinical symptoms can be caused by mutations in VGKC genes, particularly in episodic ataxia type I, which is due to a mutation in the shaker-like VGKC KCNA1, or by neurotoxins that block VGKCs, or by those that reduce inactivation of voltage-gated sodium channels (reviewed in [24]). Antibodies to VGKCs, detected by immunoprecipitation of ^{125}I-dendrotoxin VGKCs, are found in about 40% of patients with NMT and in 25% of those with CFS.

Plasma exchange and other immunotherapies are frequently effective clinically, and passive transfer of disease to mice shows evidence of peripheral nerve hyperexcitability, including repetitive action potentials in sensory nerve axons [24]. Thus, there seems little doubt that acquired NMT is an autoimmune disorder of the peripheral nervous system.

Autoimmune Associations

Like most autoimmune diseases, NMT can be associated with other conditions. In the past few years, cases coexisting with systemic sclerosis, or idiopathic thrombocytopenia, or occurring after bone marrow transfer have been reported (see [25]). NMT can also occur during penicillamine treatment. NMT can be a paraneoplastic disorder. A few reported cases have involved small-cell lung carcinoma, but up to 20% of patients may have a thymoma. In these cases it is likely, but not inevitable, that MG will be present or have been diagnosed previously. Curiously, NMT may occur in patients who have had a thymoma removed and have been successfully treated for their MG, only to start complaining of muscle cramps and fasciculation. In several patients studied in Oxford, there has been a remarkably poor correlation between antibody levels to AChR and to VGKC, suggesting not only that these two highly specific autoantibodies are driven by different immune responses, but that the T cell and B cell clones involved are subject to quite different controlling mechanisms (Fig. 2). Thus these cases of paraneoplastic NMT are similar to those of MG with thymoma, raising complex issues regarding the origin and maintenance of immune responses to these highly specific autoantigens.

NMT and the Autonomic System

Relatively little is known about the involvement of different VGKCs in the autonomic system, but patients with NMT frequently complain of excessive sweating. Although this can be attributed to the continuous muscle activity that they may suffer, it appears to be an independent phenomenon. Moreover, other autonomic symptoms can be present in patients with NMT. In these cases, central disturbance is usually also evident.

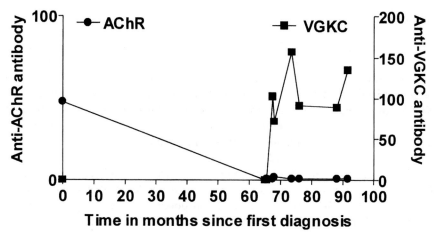

Fig. 2. Antibodies to both acetylcholine receptors and to voltage-gated potassium channels (*VGKC*) are quite common in patients with thymoma-associated myasthenia, and often show a lack of correlation. This female patient had a thymoma treated with chemotherapy. Seven years later, when her myasthenia was well under control, she developed VGKC antibodies and neuromyotonia. Clinical information from Prof. J. Newsom-Davis

CNS Symptoms in NMT

VGKC antibodies have recently been detected in patients with central nervous and autonomic nervous system involvement and limbic signs. The role of VGKC antibodies in these conditions is less clear, and their study should help to establish approaches for studying the role of antibodies to CNS antigens in disease, and ˢhed light on the complex interactions between peripheral humoral factors and ₁NS function.

Cases of NMT with central disturbance are classically referred to as Morvan's ˙ibrillary chorea or Morvan's syndrome. Typically, the florid case of Morvan's syndrome has a triad of muscle hyperactivity, autonomic symptoms including sweating and other excessive secretions and cardiac arrhythmias, and CNS disturbance including memory loss, confusion, anxiety, hallucinations and, typically – though not always reported – sleep disorders [26].

One case report illustrates this very interesting condition [27]. The patient, a 76-year-old man, presented with muscle weakness and fatigue, muscle twitching, excessive sweating and salivation, small joint pain, itching and weight loss. Over the next 12 months, he developed confusional episodes with spatial and temporal disorientation, visual and auditory hallucinations, complex behaviour during sleep and progressive nocturnal insomnia associated with diurnal drowsiness. There was also severe constipation, urinary incontinence and excessive lacrimation. On admission, he was confused and restless, and disoriented in time and space. Marked hyperhydrosis and excessive salivation and lacrimation were evident. Neurological examination disclosed continuous myokymia and myoclonus involving most of the body; there was diffuse erythema, especially on the trunk, with scratching lesions of the skin. He had frequent extrasystoles. All routine investigations were normal.

Antibodies to VGKC were detected and the patient underwent ten sessions of plasma exchange. After the third session his itching, sweating, mental disturbances and complex nocturnal behaviour improved, and these symptoms completely disappeared after the sixth session, with improvement in insomnia and reduced muscle twitching. The symptoms recurred during the following months, and he died unexpectedly during a plasma exchange treatment.

The patient had VGKC antibodies and responded to plasma exchange, clearly implicating circulating factors in the disease aetiology. Not only the peripheral NMT but also the excessive secretions and itchy skin rash improved. However, after treatment, when he was well enough to undergo more extensive testing, there were marked abnormalities of diurnal levels of neuropeptides and hormones, with raised cortisol and norepinephrine levels and reduced levels and flattened circadian rhythms of prolactin and melatonin. Particularly intriguing was the fact that many of the clinical features of Morvan's disease and the changes in hormone and peptide levels were very similar to those previously reported in fatal familial insomnia (FFI), a genetic prion disease [28]. In FFI, the main pathology appears to be in the thalamus; thus, it is possible that the VGKC antibodies in Morvan's disease alter thalamic function in a similar manner. Alternatively, they may affect the hypothalamus (where the blood-brain barrier is known to be leaky), and cause thalamic changes secondarily. A third possibility is that the very striking hypersecretion in this patient led to excess norepinephrine secretion in the periphery, with all the CNS changes resulting as a consequence of disturbed catecholamines. These possibilities need extensive investigation in order for us to try to understand the mechanisms by which VGKC antibodies in the serum can lead to such marked CNS symptoms.

VGKC Antibodies in Other Conditions

The involvement of VGKC antibodies in Morvan's syndrome led us to ask whether such antibodies were present in other patients with comparable CNS symptoms. The CNS symptoms of Morvan's syndrome are similar in many respects to those of limbic encephalitis. This condition includes anxiety, short-term memory loss and disorientation. The onset is usually subacute and the course progressive. It is frequently associated with small-cell lung cancer or other tumours, and antibodies to Hu or Ma antigens are present (see Chap. 7), and in these conditions the prognosis is usually very poor. However, in some cases it is associated with thymomas and with CV2 antibodies (see Chap. 8). In two cases of limbic encephalitis with CV2 antibodies, the patients improved when the thymomas were removed (see Chap. 8), suggesting that it can be a reversible disorder.

These observations suggested that some cases of limbic encephalitis might be associated with VGKC antibodies rather than the more typical paraneoplastic antibodies. A woman with a long history of relapsing episodes of MG, who presented 10 years after thymomectomy with a recurrence of her tumour, developed a single episode of subacute limbic encephalitis which improved substantially after plasma exchange. Retrospective analysis of her serum, taken over a 11-year period, showed a marked peak of VGKC antibodies coinciding with her limbic symptoms [28]. Another woman had no thymic or other tumour but developed subacute limbic encephalitis symptoms which resolved slowly over a period of 2 years. Again, retrospective analsyis of her serum showed a strong correlation between VGKC antibodies and limbic symptoms [29]. Interestingly, both these women had signs of excessive secretions, which are not usually a feature of limbic encephalitis. Thus the combination of secretions and limbic signs might be an indication of a VGKC-antibody-mediated form of disease which is likely to respond to plasma exchange or other immunotherapies. In addition, these limbic encephalitis patients did not have symptoms of NMT, suggesting perhaps that the specificity of their antibodies is different to those in Morvan's syndrome and acquired neuromyotonia. VGKCs are highly pleomorphic, being formed as tetramers of variable proportions of up to six different gene products (KCNA1-6); thus, specificity for different oligomers might underlie some of the differences in clinical expression of diseases associated with VGKC antibodies. Alternatively, it is possible that the peripheral tissue can upregulate other VGKCs or in some other way compensate for their loss.

Conclusions

The study of the peripheral autoimmune conditions has been an essential step in defining the importance of specific ion channels in normal function, the mechanisms of action of antibodies in causing ion channel loss, and the paradigms for defining an antibody-mediated disease (see Introduction to this volume). We may now be starting a new era when these essential observations can be used to help understand a wide range of other peripheral diseases, CNS disorders, and even developmental conditions. These conditions are discussed in a little more detail in [30].

References

1. Sanders DB, Andrews I, Howard JF, Massey JM (1997) Seronegative myasthenia gravis. Neurology 48:S40-S45
2. Mossman S, Vincent A, Newsom-Davis J (1986) Myasthenia gravis without acetylcholine-receptor antibody: a distinct disease entity. Lancet 1:116-119
3. Blaes F, Beeson D, Plested P et al (2000) IgG from "seronegative" myasthenia gravis patients binds to a muscle cell line, TE671, but not to human acetylcholine receptor. Ann Neurol 47:504-510
4. Brooks EB, Pachner AR, Drachman DB, Kantor FS (1990) A sensitive rosetting assay

for detection of acetylcholine receptor antibodies using BC3H-1 cells: positive results in 'antibody-negative' myasthenia gravis. J Neuroimmunol 28:83-93

5. Yamamoto T et al (1991) Seronegative myasthenia gravis: a plasma factor inhibiting agonist-induced acetylcholine receptor function copurifies with IgM. Ann Neurol 30:550-557

6. Plested CP, Newsom-Davis J, Vincent A (1998) Seronegative myasthenia plasmas and non-IgG fractions transiently inhibit nAChR function. Ann New York Acad Sci 841:501-504

7. Plested CP (1999) Mechanism of action of seronegative myasthenia. DPhil thesis, University of Oxford

8. Hoch W (1999) Formation of the neuromuscular junction: agrin and its unusual receptors. Eur J Biochem 265:1-10

9. Hopf C, Hoch W (1998) Dimerization of the muscle-specific kinase induces tyrosine phosphorylation of acetylcholine receptors and their aggregation on the surface of myotubes. J Biol Chem 273:6467-6473

10. Hoch W, McConville J, Helms S et al (2001) Autoantibodies to the receptor tyrosine kinase MuSK in patients with myasthenia gravis without acetylcholine receptor antibodies. Nat Med 7:365-368

11. Hall JG, Vincent A (2001) Arthrogryposis. In: Jones HR, De Vivo DC, Darras BT (eds) Neuromuscular disorders of infancy and childwood. Butterworth-Heinemann, Woburn, USA (in press)

12. Drachman DB, Coulombre A (1962) Experimental clubfoot and arthrogryposis multiplex congenita. Lancet ii:523-526

13. Moessinger A (1983) Fetal akinesia deformation sequence: an animal model. Pediatrics 72:857-863

14. Venet-der Garabedian B, Lacokova M, Eymard B et al (1994) Association of neonatal myasthenia gravis with antibodies against the foetal acetylcholine receptor. J Clin Invest 94:555-559

15. Riemersma S, Vincent A, Beeson D et al (1997) Association of arthrogryposis multiplex congenita with maternal antibodies inhibiting fetal acetylcholine receptor function. J Clin Invest 98:2358-2363

16. Brueton LA, Huson SM, Cox PM et al (2000) Asymptomatic maternal myasthenia as a cause of the Pena-Shokeir phenotype. Am J Med Genet 1:92:1-6

17. Carr SR, Gilchrist JM, Abuelo DN, Clark D (1991) Treatment of antenatal myasthenia gravis. Obstet Gynecol 78:485-489

18. Polizzi A, Huson SM, Vincent A (2000) Teratogen update: maternal myasthenia gravis as a cause of congenital arthrogryposis. Teratology 62:332-341

19. Jacobson L, Beeson D, Tzartos S, Vincent A (1999) Monoclonal antibodies raised against human acetylcholine receptor bind to all five subunits of the fetal isoform. J Neuroimmunol 98:112-120

20. Vincent A, Matthews I, Newsom-Davis J, Willcox N (2000) Antibodies to fetal acetylcholine receptors in parous women. Ann Neurol 48:479 (abs)

21. Bianchi DW (2000) Fetomaternal cell trafficking: a new cause of disease? Am J Med Genet Mar 6:91:22-28

22. Pinto A, Gillard S, Moss F et al (1998) Human autoantibodies specific for the a1A calcium channel subunit reduce both P-type and Q-type calcium currents in cerebellar neurons. Proc Natl Acad Sci USA 95:8328-8333

23. Trivedi R, Mundanthanam G, Amyes E et al (2000) Autoantibody screening in subacute cerebellar ataxia. Lancet 356:565-566

24. Hart IK (2000) Acquired neuromyotonia: a new autoantibody-mediated neuronal potassium channelopathy. Am J Med Sci 319:209-216
25. Vincent A (2000) Understanding neuromyotonia. Muscle Nerve 23:655-657
26. Serratrice G, Azulay JP (1994) Mise au point. Que reste-t-il de la chorée fibrillaire de Morvan? Rev Neurol 150:257-65
27. Liguori R, Vincent A, Clover L et al (2001) Morvan's syndrome: peripheral and central nervous and cardiac involvement with antibodies to voltage-gated potassium channels and altered circadian rhythyms. Brain (*in press*)
28. Lugaresi E, Medori R, Montagna P et al (1986) Fatal familial insomnia and dysautonomia with selective degeneration of thalamic nuclei. N Engl J Med 315: 997-1003
29. Buckley C, Oger J, Clover L et al (2001) Potassium channel antibodies in two patients with reversible limbic encephalitis. Ann Neurol 50:74-79
30. Vincent A, Lily O, Palace J (1999) Pathogenic autoantibodies to neuronal proteins in neurological disorders. J Neuroimmunol 100:169-180

Chapter 2

Pregnancy and Myasthenia Gravis

A.P. Batocchi

Immunology of Pregnancy

Pregnancy is an immunological balancing act in which the mother's immune system has to remain tolerant of paternal major histocompatibility (MHC) antigens while maintaining normal immune competence for defence against microorganisms.

Fetal trophoblast plays a major role in evading recognition by the maternal immune system. Trophoblast cells fail to express MHC class I and class II molecules and express Fas ligand that induces apoptosis of any Fas-expressing maternal immune cells at the placenta/decidua interface [1]. However, there is evidence that the maternal immmune response can influence growth and survival of the fetoplacental unit by local cytokine production. Uterine decidual and placental cells spontaneously secrete Thelper 2 (Th2) type cytokines such as interleukin 4 (IL-4), IL-5, IL-10, type I interferons (IFN) and a transforming growth factor (TGF) β_2-like cytokine during all three trimesters of pregnancy [2, 3]. Conversely, the inflammatory cytokines IL-2, tumor necrosis factor-α (TNFα) and IFNγ terminate normal pregnancy when injected into pregnant mice [3], whereas granulocyte/macrophage colony stimulating factor (GM-CSF), IL-3 and IL-10 will protect against fetal resorption [2].

During normal pregnancy there is a down-regulation of the cellular immune response with a shift in the balance of cytokine profile away from Th1 type reactivity to a Th2 type reactivity, both in the maternal peripheral blood and in the fetomaternal unit. Th2 cytokines predominate in the early pregnant decidua while Th1 cytokines such as IFNγ, IL-2, IL-12, TNFβ predominate in the non-pregnant endometrium [4]. Moreover, they are higher in non-pregnant women with recurrent miscarriages than in fertile non-pregnant women [5]. In the peripheral blood, cytotoxic effects of T cells are reduced in normal pregnancy, and both CD4+ and CD8+ T cells produce fewer Th1 cytokines and more Th2 cytokines, especially in the second and third trimester and shortly after delivery [6, 7]. There is also increased immunoglobulin (Ig) production with a deflection from a cytotoxic IgG2 (Th1-induced) to a non-cytotoxic IgG1 (Th2-induced) antibody response [3]. Peripheral blood mononuclear cells from healthy fertile women

Institute of Neurology, Catholic University, L.go F. Vito 1, 00168 Rome, Italy.
e-mail: annapaola_b@hotmail.com

Table 1. Cytokines and pregnancy

	Pregnancy	Recurrent Miscarriages
Feto-placental unit	↓ Cellular immunity	↑ Cellular immunity
	↑ Th2 cytokines (IL-4, IL-3, IL-5, IL-10)	↑ Th1 cytokines (IL-2, IL-12, TNFα, IFNγ)
	↑ Type1 IFN, TGFβ 2-like cytokine, GM-CSF	↑ Natural killer
Maternal peripheral blood	↓ Cellular immunity	↑ Cellular immunity
	↑ Th2 cytokines	↑ Th1 cytokines
	↑ Ig, IgG1, IgG2	

respond in vitro to trophoblast antigens by producing IL-10, while most women who have frequent miscarriages respond by producing embryotoxic activity and high levels of IFNγ and TNFβ [3]. These last data suggest that Th2 type immunity may be a natural response to trophoblast antigens that is necessary for a successful pregnancy [2]. The observations are summarised in Table 1.

Several cells and soluble factors have been proposed as candidates for mediators of the Th1- to-Th2 shift during pregnancy. Factors secreted by cultured placental cells (placental suppressor factor), trophoblast cell lines (trophoblast cell-derived factor), a progesterone-induced blocking factor (PIBT), progesterone itself [2] and some oestrogens [8] appear to favour the development of human Th2 type cells but their roles in pregnancy are still not clear.

Pregnancy and Infectious and Autoimmune Diseases

Clinical studies suggest that immunomodulatory effects of pregnancy at the cytokine level, consistent with a weakening of cell-mediated immunity and a strengthening of humoral immunity, correlate with changes in the clinical course of infectious and autoimmune diseases during gestation [3].

Infectious diseases caused by intracellular pathogens such as HIV, leprosy, malaria, toxoplasmosis or tuberculosis appear to be exacerbated by pregnancy. For instance, systemic lupus erythematosus, an antibody-mediated disorder, tends to flare up during gestation; while most women affected by cell-mediated disorders such as rheumatoid arthritis, multiple sclerosis, autoimmune uveitis or autoimmune thyroid diseases experience a temporary remission of symptoms [3]. The clinical course of the mouse model of multiple sclerosis, experimental autoimmune encephalomyelitis (EAE), is also less severe during late pregnancy, and the hormone responsible for this decrease in severity seems to be oestriol, the primary oestrogen produced by the fetoplacental unit during pregnancy [8]. Ovariectomy does not appear to have an influence on the susceptibility to or severity of experimental autoimmune myasthenia gravis (EAMG) [9].

All autoimmune diseases tend to become exacerbated in the post-partum period. Prolactin, the lactogenic hormone, is considered a natural immune enhancer, and mild hyperprolactinaemia seems to be a risk factor for the development or deterioration of autoimmunity [10].

Myasthenia Gravis and Cytokines

Myasthenia gravis (MG) is an organ-specific autoimmune disease affecting the neuromuscular junction. The disorder results in the loss of functional acetylcholine receptors (AChR) at the postsynaptic membrane and is mediated by autoantibodies and complement. Serum antibodies against AChR are found in most myasthenic patients and their production is mediated by cytokines produced by T cells [11]. About 10% of patients do not possess circulating anti-AChR antibodies and might have autoantibodies to other components of the motor endplate [12].

The pathogenetic mechanisms of the disease are not completely understood. Both AChR-specific IFNγ, IL-2-secreting Th1 cells and IL-4, IL-10-secreting Th2 cells are present in the peripheral blood of MG patients [13], but it is not yet clear which are essential for driving the production of pathogenic anti-AChR antibodies. EAMG is a T-cell-dependent, antibody-mediated autoimmune disease induced in animals by a single immunization with AChR, and represents an animal model of the human disease. Recent studies show that a Th1 response is required for the development of EAMG in mice [14], as IFNγ knockout mice are resistant to EAMG [15] while IL-4 knockout mice readily develop the disease [16]. On the other hand, Th1 responses do not appear to be crucial for the development of rat EAMG, since in that case complement-fixing anti-AChR antibodies can be generated as a consequence of either Th1- or TH2-mediated T cell help [17]. Further studies on human MG are required to clarify the role of cytokines in the pathogenesis of the disease.

MG and Pregnancy

MG is more common in women than in men in the second and third decades of life, overlapping with childbearing years [18], but there are very few studies on the influence of pregnancy on the course of MG and on the potential effect of the disease on pregnancy outcome and fetal growth [19-21].

Pregnancy does not worsen the long-term outcome of MG. The course of the disease seems to be highly variable and unpredictable during gestation. Approximately one-third of patients deteriorate, one-third remain unchanged, and one-third improve (Table 2). There is no correlation between MG severity before conception, and exacerbation of symptoms during pregnancy. The disease can recur during gestation in women who are in full remission without any therapy before conception [21]. The clinical course of the disease during one pregnancy does not predict the course during a subsequent pregnancy [21]. The first trimester and the post-partum period seem to be the most critical periods for MG exacerbation [19, 21].

Table 2. Pregnancy effects on MG course (408 pregnancies in 304 women[#])

Pregnancy		
No change	142	(35%)
Improvement	109	(27%)
Worsening	157	(38%)
After delivery		
Worsening	126	(28%)
Maternal deaths	9	(2%)

[#] Data from Plauché [19], Eymard [20], Batocchi [21]

Unlike other antibody-mediated autoimmune diseases such as systemic lupus erythematosus, which carries a clearly increased risk of an acute flare-up during pregnancy [22], the high variability of MG during gestation, even in the same patient, does not help to clarify the role of different cytokine patterns in the development of the disease and suggests that both Th1 and Th2 cells may, indeed, be involved in the pathogenesis of MG.

With regard to the influence of MG on pregnancy outcome, the incidence of spontaneous abortion is very low in MG patients (see Table 3) and pregnancy termination does not seems to influence the disease outcome [19, 21]. An increased prevalence of premature labour has been reported in one study [19] but not confirmed in another [21] .

The prevalence of caesarean section is not statistically different from that of the general population (Tab. 3). Vaginal delivery with epidural anaesthesia to reduce pain and fatigue is preferred in patients with mild MG signs, and caesare-

Table 3. MG effects on pregnancy outcome[#]

	MG	Italy[°]	p
N° Pregnancies	386		
Abortion	48		
Spontaneous	17 (43*)	(100*)	< 0.001
Ectopic pregnancy	1		
Voluntary induced	30		
N° deliveries	106		
Method of delivery			
Vaginal	83		
Caesarian section	23 (21%)	(24%)	n.s.
Duration of pregnancy			
Term	83		
Premature	23 (21%)		

* x 1000 live births; n.s., not significant
[#] Data from Plauché [19], Eymard [20], Batocchi [21]
[°] ISTAT, Annuario Statistico Italiano 1994-1997

an section should be performed only for clear obstetric indications [19]. General endotracheal anaesthesia for caesarean delivery is recommended only in patients with bulbar or respiratory involvement [19].

The risks to the fetus include prematurity, neonatal myasthenia and arthrogryposis multiplex congenita. An increased prevalence of low birth weight and perinatal death due to prematurity, babies small for their gestational age, neonatal myasthenia and arthrogryposis multiplex congenita have been reported [19].

Neonatal MG

Neonatal MG (NMG) is a transient disease affecting about 10-20% of the infants of myasthenic mothers [19, 21, 23-25]. An unusually high incidence of NMG, about 50%, has been reported in France [20]. The disease is due to the transplacental transmission of anti-AChR antibodies from mother to fetus and represents a human model of passively transferred autoimmune disease. As maternal IgG crosses the placenta throughout the last two trimesters [26], its pathogenic effect may be achieved before birth. The birth of an affected child increases the chance that subsequent children will be similarly affected [27].

NMG has been reported also in babies born to anti-AChR-seronegative myasthenic mothers [28] and may be caused by the passive transfer of pathogenic autoantibodies directed towards antigens of the neuromuscular junction different from AChR.

Approximately 80% of affected children develop symptoms during the first 24 h of life, but the condition can start up to 4 days after birth. The clinical signs are weak sucking and swallowing, hypotonia and weak movements, feeble cry and respiratory difficulties; ptosis and facial weakness are less prominent [19, 25]. It is necessary to observe carefully the newborn of every myasthenic mother for signs of weakness of skeletal muscles, particularly those that control breathing and swallowing, for several days before the baby can be discharged. The symptoms may last from a few days up to 15 weeks [19, 25], but a long-lasting evolution has been reported [20]. In an unusual case, impairment of swallowing and ophthalmoparesis lasted 13 months and facial weakness persisted [20]. The diagnosis of NMG is made by the presence of detectable serum anti-AChR antibodies, abnormal decrement (>11%) of the third to fifth compound muscle action potentials (CMAP) on low-rate supramaximal repetitive nerve stimulation, and temporary improvement of strength following an injection of neostigmine (usually 0.1 mg i.m.) or edrophonium (usually 1 mg) [18]. The prognosis is usually good: the affected babies recover completely and do not develop MG later in life. Management consists of anticholinesterases and careful nursing (suction of pharyngeal secretion, tube feeding) and artificial respiration if needed. In the most severe cases exchange transfusion can be performed. High-dose intravenous immunoglobulin has been used in some cases, with discordant results [29, 30].

There is no correlation between the occurrence of neonatal MG and the severity of maternal MG during pregnancy: mothers in remission can give birth to affected infants, while the newborn of a severely affected mother can show no sign of NMG [19, 21]. However, immunosuppressive therapy may prevent NMG, since the newborns of severely affected patients treated during pregnancy with immunosuppressive therapy showed no signs of NMG, while the babies of severely affected women who had only received anticholinesterases suffered from NMG [21].

Not all babies born to seropositive MG patients have detectable anti-AChR antibodies, and not all babies with detectable anti-AChR antibodies show signs of NMG [21, 24]. Although it has been reported that anti-AChR titres are higher in "transmitting" than in "non-transmitting" mothers, and correspondingly higher in affected than in asymptomatic newborns [31], most authors found a poor correlation between the level of maternal anti-AChR antibodies and the occurrence or the severity of NMG [21, 23, 24, 31]. The disease is more frequent in seropositive babies, but affected babies do not uniformly have a higher anti-AChR antibody level than healthy ones, and there is no correlation between newborn anti-AChR antibody titre and severity of NMG. However, some babies show a higher antibody titre than their mothers [21, 24, 32].

Why NMG develops in only 10-20% of babies born to MG mothers and why most seropositive babies do not display signs of MG is still unclear. Lefvert and Osterman [24] found a marked difference between anti-AChR antibody idiotypes in mothers and in affected children, suggesting that a transient synthesis of anti-AChR antibodies in the children might be a factor in the pathogenesis of NMG. This hypothesis has not been confirmed by other authors, who pointed out similar antibody specificities in sera from mothers and infants [31, 33], and no difference in antibody profile between mothers who transferred MG and those who did not [33]. Other studies support the hypothesis that other factors in the fetal environment may also contribute to the clinical manifestation of NMG [31, 34]. Brenner et al. [34] demonstrated that α-fetoprotein blocks the binding of anti-AChR antibodies to AChR in vitro and suggested that the high level of α-fetoprotein in a fetus and in a pregnant woman in the second trimester might provide non-specific protection to mother and infant. Another work, however, has failed to confirm the inhibitory effect of α-fetoprotein [24]. More recent studies [32] showed a close relation between the occurrence of NMG and the ratio of maternal antibodies directed at the fetal or adult isoforms of the AChR, suggesting that autoantibodies directed against fetal AChR could play a predominant role in the pathogenesis of NMG. A high anti-fetal/anti-adult anti-AChR antibody titre ratio in a mother at any time, before or during pregnancy, seems to be predictive of the occurrence of NMG in a first child. It has not yet been established whether that is also true for subsequent pregnancies.

Arthrogriposis Multiplex Congenita

Arthrogriposis multiplex congenita (AMC) is a congenital disorder characterized by multiple joint contractures, in some cases associated with other abnormalities, which result from lack of fetal movements in utero. Some cases are genetically determined but infectious and physical agents, drugs, toxins and maternal neurological and muscular diseases associated with reduced fetal movements have been implicated in the pathogenesis of the disease [35]. Several cases of AMC associated with NMG have been described [35-37], but there are no data on the overall prevalence of AMC in infants of myasthenic mothers. Literature review suggests that the disease is not so common, but there is a high risk of recurrence in maternal MG-associated AMC [36]. Riemersma et al. [37] showed that antibodies that inhibit the function of fetal AChR are present in some anti-AChR-seronegative mothers with no signs of MG, and in seropositive asymptomatic or symptomatic myasthenic mothers with an obstetric history of recurrent AMC. Injection into pregnant mice of plasma from seronegative or seropositive mothers with anti-fetal AChR function activity caused paralysis and AMC in mouse pups [35]. These data show the pathogenic role of maternal antibodies to fetal antigens in some cases of AMC, but it is still not clear whether the disease is caused by the anti-AChR antibody itself or by different autoantibodies against other fetal antigens.

Treatment of Myasthenia Gravis in Pregnancy

MG is a chronic autoimmune disease and treatment is frequently necessary before, during and after pregnancy in MG patients to ensure maternal and fetal well-being.

Anticholinesterase (AChE) drugs have been reported to be safe during pregnancy. A single case of intestinal tube muscular hypertrophy in the newborn of a myasthenic patient treated during pregnancy with very high doses of AChE has been described [38]. Corticosteroids, azathioprine and, in some cases, cyclosporine may be needed when myasthenic symptoms are not satisfactorily controlled with AChE.

Knowledge of the potential effect of the use of immunosuppressive drugs in myasthenic patients during pregnancy is limited, rendering decision-making difficult for both patient and physician. The safety of corticosteroids, azathioprine, cyclosporine and tacrolimus is relatively well known in pregnancies of transplant recipients [39-43], and in patients suffering from other autoimmune diseases such as systemic lupus erythematosus [44, 45] and inflammatory bowel disease [46]. Successful pregnancies are reported in most kidney [39, 40], liver [41] and heart [42] transplant recipients, not only those treated with costicosteroids and azathioprine, but also with cyclosporine. No fetal anomalies have been observed [39-43], but patients receiving cyclosporine experienced a higher frequency of spontaneous abortions and preterm deliveries than the women who received corticosteroids and azathioprine [39]. Adverse pregnancy outcomes and fetal or neonatal loss were less frequent in systemic lupus erythematosus patients treated with immunosuppressive drugs than in untreated ones [44] and no macroscopic malformations were found in live children of women who took azathioprine during pregnancy [45]. There were no congenital abnormalities or subsequent health problems in children of women affected by inflammatory bowel disease treated with azathioprine during pregnancy [46].

Fetal malformations may be associated with maternal use of methotrexate (MTX). MTX and aminopterin have been used to induce abortions, and congenital abnormalities resulting from unsuccessful abortion have been referred to as "aminopterin syndrome" [47]. This syndrome is characterized by central nervous system, skeletal and cardiac abnormalities. We recommend that MTX not be used to treat MG women of childbearing age.

Treatment with corticosteroids presents little, if any, teratogenic risk to the fetus; only a slight increase in the incidence of cleft palate has been reported [48]. Premature rupture of the membranes may be related to high-dose corticosteroid treatment [49]. Whether this therapy increases either neonatal or maternal infection is still unclear [50].

Azathioprine crosses the placenta, and at high doses there is some evidence of teratogenic effects in animals [51]. Although there are several reports of adverse effects of azathioprine on children of mothers treated with this drug, including teratogenicity, chromosomal effects, immunological and haematological depression [46, 52], low birth weight and abortion [45], there has never been any definite demonstration of teratogenicity due to azathioprine in humans at therapeutic doses. More recent clinical studies confirm that there is no increase in the rate of fetal malformation, abortion or prematurity when azathioprine is administered at therapeutic doses, even in early pregnancy [46].

Previous reports have not shown any teratogenicity due to cyclosporine [39-42] or tacrolimus [43] in transplant recipients. In these patients, drugs seem to carry a higher risk of spontaneous abortion, prematurity, babies small for their gestational age [39-43] and temporary elevation of serum creatinine concentration in infants, as well as hyperkalaemia [43], but the contribution of the underlying maternal disease and the presence of renal dysfunction at conception [53] should also be considered.

Patients taking azathioprine or cyclosporine who wish to conceive should be fully informed about both the fetal risks and the possibility of MG worsening after drug withdrawal. It has been reported [21] that sudden suspension of azathioprine had no effect on the MG course in one patient but induced a severe exacerbation of MG symptoms in another. Although some clinicians [54] recommend that azathioprine and cyclosporine should always be withdrawn in pregnancy, in our opinion, drugs can be continued throughout pregnancy to avoid life-threatening exacerbation if they play an important role in controlling the disease. However, the final decision on stopping or continuing to take the drug should be taken by the women concerned.

Plasmapheresis [21, 55] and high-dose gammaglobulins [21, 56] are effective and safe treatments for myasthenic crises also in pregnancy.

Treatment of MG During Breast-Feeding

Anti-AChR antibody has been found in human colostrum and breast milk of myasthenic women with high antibody levels [57]. Thus, breast-feeding could supply additional antibodies to the baby, but the lack of correlation between anti-AChR titre and the occurrence of NMG precludes breast-feeding only for mothers of affected children.

Corticosteroids, azathioprine, cyclosporine, MTX and AChE drugs are transferred to breast milk at low levels [52]. Corticosteroids can be used safely during lactation, while breast-feeding is contraindicated in mothers receiving other immunosuppressive medication as they may induce immunosuppression in the babies. Large doses of AChE drugs may preclude breast-feeding because this medication could cause gastrointestinal upsets in the breast-fed newborn [58].

Conclusions

Maternal immune response can influence the growth and survival of the fetus. To ensure a successful pregnancy, during gestation a down-regulation of the cellular immune response occurs, with a shift in the balance of cytokine profile away from Th1 type to Th2 type reactivity, both in the fetomaternal unit and in the maternal peripheral blood. Clinical studies suggest that this immunoregulatory effect at the cytokine level can influence the clinical course of autoimmune diseases during pregnancy: antibody-mediated diseases tend to flareup while cell-mediated disorders tend to remit.

MG is an organ-specific autoimmune disorder caused by anti-AChR antibodies, but both Th1 and Th2 cells seem to be involved in the pathogenesis of the disease. The course of MG during gestation is highly variable and unpredictable and may change in subsequent pregnancies. Pregnancy does not worsen the long-term outcome of the disease. The incidence of spontaneous abortion is very low in MG patients, but the risks to the fetus include prematurity, NMG and, rarely, AMC. NMG is a transient disease affecting about 10%-20% of the infants of mothers with MG due to the transplacental transmission of anti-AChR antibodies. The occurrence of NMG does not correlate either with maternal disease severity or maternal anti-AChR antibody titre.

Corticosteroids, plasmapheresis and intravenous human immunoglobulins can be safely used to treat MG patients during gestation, while azathioprine and cyclosporine can be used with caution if they are indispensable to control the disease. Anti-AChR antibodies are present in the milk of myasthenic women, but the lack of correlation between antibody titre and the occurrence of NMG precludes breast-feeding only in mothers of affected babies or those receiving immunosuppressive therapy.

Very close cooperation between neurologist, obstetrician and neonatologist is recommended for good management of pregnant MG patients.

References

1. Weetman AP (1999) The immunology of pregnancy. Thyroid 9:643-646
2. Wegmann TG, Lin H, Guilbert L, Mosmann TR (1993) Bidirectional cytokine interactions in the maternal-fetal relationship: is successful pregnancy a Th_2 phenomenon? Immunol Today 14:353-356
3. Raghupathy R (1997) Th1-type immunity is incompatible with successful pregnancy. Immunol Today 18:478-482
4. Saito S, Tsukaguchi N, Hasegawa T et al (1999) Distribution of Th1, Th2, and Th0 and the Th1/Th2 cell ratios in human peripheral and endometrial T cells. Am J Reprod Immunol 42:240-245
5. Lim KJH, Odukoya OA, Ajjan RA et al (2000) The role of T-helper cytokines in human reproduction. Fertility Sterility 73:136-142
6. Reinhard G, Noll A, Schlebusch H et al (1998) Shifts in the TH1/TH2 balance during human pregnancy correlate with apoptotic changes. Biochem Biophys Res Commun 245:933-988
7. Saito S, Sakai M, Tanebe K et al (1999) Quantitative analysis of peripheral blood Th0, Th1, Th2 and the Th1:Th2 cell ratio during normal human pregnancy and preeclampsia. Clin Exp Immunol 117:550-555
8. Kim S, Liva SM, Dalal MA et al (1999) Estriol ameliorates autoimmune demyelinating disease. Implications for multiple sclerosis. Neurology 52:1230-1238
9. Leker RR, Karni A, Brenner T et al (2000) Effects of sex hormones on experimental autoimmune myasthenia gravis. Eur J Neurol 7:203-206
10. Neidhart M (1998) Prolactin in autoimmune diseases Proc Soc Exp Biol Med 217:408-419
11. Vincent A, Newson-Davis J (1985) Acetylcholine receptor antibody as a diagnostic test for myasthenia gravis: results in 153 validated cases and 2967 diagnostic assays. J Neurol Neurosurg Psychiatry 48:1246-1252
12. Evoli A, Batocchi AP, Lo Monaco M et al (1996) Clinical heterogeneity of seronegative myasthenia gravis. Neuromusc Disord 6:155-161
13. Zhang G, Navikas V, Link H (1997) Cytokines and the pathogenesis of myasthenia gravis. Muscle Nerve 20:543-551
14. Balasa B, Sarvetnick N (2000) Is pathogenic humoral autoimmunity a Th1 response? Lessons from (for) myasthenia gravis. Immunol Today 21:19-23
15. Balasa B, Deng C, Lee J et al (1997) Interferon gamma (IFN-gamma) is necessary for the genesis of acetylcholine receptor-induced clinical experimental autoimmune myasthenia gravis in mice. J Exp Med 4:385-391
16. Balasa B, Deng C, Lee J et al (1998) The Th2 cytokine IL-4 is not required for the progression of antibody-dependent autoimmune myasthenia gravis. J Immunol 161:2856-2862
17. Saoudi A, Bernard I, Hoedemaekers A et al (1999) Experimental autoimmune myasthenia gravis may occur in the context of a polarized Th1- or Th2-type immune response in rats. J Immunol 162:7189-7197
18. Oosterhuis HJGH (1997) Myasthenia gravis. Groningen Neurological Press, Groningen
19. Plauché WC (1991) Myasthenia gravis in mothers and their newborns. Clin Obstet Gynecol 34:82-99
20. Eymard B, Morel E, Dulac O et al (1989) Myasthénie et grossesse: une étude clinique et immunologique de 42 cas. Rev Neurol (Paris) 145:696-701
21. Batocchi AP, Majolini L, Evoli A et al (1999) Course and treatment of myasthenia gravis during pregnancy. Neurology 52:447-452

22. Wechsler B, Le Thi Huong D, Piette JC (1999) Pregnancy and systemic lupus erythematosus. Ann Med Interne (Paris) 150:408-418
23. Donaldson JO, Penn AS, Lisak RP et al (1981) Antiacetylcholine receptor antibody in neonatal myasthenia gravis. Am J Dis Child 135:222-225
24. Lefvert AK, Osterman PO (1983) Newborn infants to myasthenic mothers: a clinical study and an investigation of acetylcholine receptor antibodies in 17 children. Neurology 33:133-138
25. Giacoia GP, Azubuike K (1991) Autoimmune diseases in pregnancy: their effect on the fetus and newborn. Obstet Gynecol Surv 46:723-732
26. Harris RE (1978) Maternal and fetal immunology. Obstet Gynecol 51:733-739
27. Camus M, Clouard C (1989) Myasthénie et grossesse: à propos de 8 cas. J Gynecol Obstet Biol Reprod 18:904-911
28. Melber D (1988) Maternal-fetal transmission of myasthenia gravis with negative acetylcholine receptor antibody. N Engl J Med 318:996
29. Bassan H, Muhlbauer B, Tomer A, Spirer Z (1998) High dose intravenous immunoglobulin in the management of transient neonatal myasthenia gravis. Pediatr Neurol 18:181-183
30. Tagher RJ, Baumann R, Desai N (1999) Failure of intravenously administered immunoglobulin in the treatment of neonatal myasthenia gravis. J Pediatr 134:233-235
31. Bartoccioni E, Evoli A, Casali C et al (1986) Neonatal myasthenia gravis: clinical and immunological study of seven mothers and their newborn infants. J Neuroimmunol 12:155-161
32. Gardnerova M, Eymard B, Morel E et al (1997) The fetal/adult acetylcholine receptor antibody ratio in mothers with myasthenia gravis as a marker for transfer of the disease to the newborn. Neurology 48:50-54
33. Tzartos SJ, Efthimiadis E, Morel B et al (1990) Neonatal myasthenia gravis: antigenic specificities of antibodies in sera from mothers and their infants. Clin Exp Immunol 80:376
34. Brenner T, Beyth Y, Abramsky O (1980) Inhibitory effect of a-fetoprotein on the binding of myasthenia gravis antibody to acetylcholine receptor. Proc Nat Acad Sci USA 77:3635-3639
35. Jacobson L, Polizzi A, Morriss-Kay G, Vincent A (1999) Plasma from human mothers of fetuses with severe arthrogriposis multiplex congenita causes deformities in mice. J Clin Invest 103:1031-1038
36. Barnes RJ, Kanabar DJ, Brueton L et al (1995) Recurrent congenital arthrogriposis leading to a diagnosis of myasthenia gravis in an initially symptomatic mother. Neuromusc Disord 5:59-65
37. Riemersma S, Vincent A, Beeson D et al (1996) Association of arthrogryposis multiplex congenita with maternal antibodies inhibiting fetal acetylcholine receptor function. J Clin Invest 98:2358-2363
38. Cooker J, Thompson RM (1966) Multiple smooth muscle hypertrophies in a new born infant. Arch Dis Child 41:514-518
39. Haugen G, Fauchald P, Sodal G et al (1994) Pregnancy outcome in renal allograft recipients in Norway. Acta Obstet Gynecol Scand 73:541-546
40. Haugen G, Fauchald P, Sodal G et al (1991) Pregnancy outcome in renal allograft recipients: influence of cyclosporin A. Eur J Obstet Gynecol Reprod Biol 39:25-29
41. Scantlebury V, Gordon R, Tzakis A et al (1990) Childbearing after liver transplantation. Transplantation 49:317-321
42. Wagoner LE, Taylor DO, Olsen SL et al (1993) Immunosuppressive therapy, management, and outcome of heart transplant recipients during pregnancy. J Heart Lung Transplant 12:993-999

43. Jain A, Venkataramanan R, Fung JJ et al (1997) Pregnancy after liver transplantation under tacrolimus. Transplantation 64:559-565

44. Ramsey-Goldman R, Mientus JM, Kutzer JE et al (1993) Pregnancy outcome in women with systemic lupus erythematosus treated with immunosuppressive drugs. J Rheumatol 20:1152-1157

45. Martinez-Rueda JO, Arce-Salinas CA, Kraus A et al (1996) Factors associated with fetal losses in severe systemic lupus erythematosus. Lupus 5:113-119.

46. Alstead EM, Ritchie JK, Lennard-Jones JE et al (1990) Safety of azathioprine in pregnancy in inflammatory bowel disease. Gastroenterology 99:443-446

47. Buckley LM, Bullaboy CA, Leichtman L, Marquez M (1997) Multiple congenital anomalies associated with weekly low-dose methotrexate treatment of the mother. Arthritis Rheum 40:971-973

48. Fraser FC, Sajoo A (1995) Teratogenic potential of corticosteroids in humans. Teratology 51:45-46

49. Gaudier FL, Santiago-Delin E, Rivera J , Gonzales Z (1988) Pregnancy after renal transplantation. Surg Gynecol Obstet 167:533-543

50. Schmidt PL, Sims ME, Strassner HT et al (1984) Effect of antepartum glucocorticoid administration upon neonatal respiratory distress syndrome and perinatal infection. Am J Obstet Gynecol 178:178-186

51. Tuchmann-Duplessis H, Mercier-Parot L (1996) Production in rabbits of malformations of the limbs by azathioprine and 6-mercaptopurine. CR Soc Biol 166:501-506

52. Ramsey-Goldmman R, Schilling E (1997) Immunosuppressive drug use in pregnancy. Rheum Dis Clin North Am 23:149-167

53. Casele HL, Laifer SA (1998) Association of pregnancy complications and choice of immunosuppressant in liver transplant patients. Transplantation 27:581-583

54. Present DH, Meltzer SJ, Krumholz MP et al (1989) 6-mercaptopurine in the managment of inflammatory bowel disease: short and long term toxicity. Ann Intern Med 111:641-649

55. Watson WJ, Katz VL, Bowes WA (1984) Plasmapheresis during pregnancy. Obstet Gynecol 76:451-457

56. Kaaja R, Julkunen A, Ammala P et al (1993) Intravenous immunoglobulin treatment of pregnant patients with recurrent pregnancy losses associated with antiphospholipid antibodies. Acta Obstet Gynecol Scand 72:63-66

57. Brenner T, Shahin R, Steiner I , Abramsky O (1992) Presence of anti-acetylcholine receptor antibodies in human milk: possible correlation with neonatal myasthenia gravis. Autoimmunity 12:315-316

58. Osserman KE, Kornfild P, Cohen E et al (1958) Studies in myasthenia gravis. AMA Arch Intern Med 102:72-81

Chapter 3

Immunological Mechanisms in the Lambert-Eaton Myasthenic Syndrome

A. EVOLI, P.A. TONALI

The Lambert-Eaton myasthenic syndrome (LEMS) is an antibody-mediated disorder of neuromuscular and autonomic synaptic transmission. It is paraneoplastic in about 50% of cases, generally being associated with small-cell lung carcinoma (SCLC). In non-cancer cases, it can be associated with other autoimmune diseases [1].

Pathophysiology

Although first reported by Anderson in 1953 [2], LEMS was first clearly distinguished from myasthenia gravis (MG) by Eaton and Lambert who, in two papers dated 1957 and 1962 [3, 4], described the characteristic electromyographic features which were suggestive of a presynaptic disorder of neuromuscular transmission. In 1971, Lambert and Elmquist showed clearly by electrophysiologic studies on patients' muscle biopsies [5] that the neuromuscular transmission defect in LEMS is presynaptic and consists of a decreased quantal content (i.e. the number of quanta or packets of acetylcholine released by each nerve impulse). On the other hand, the miniature end-plate potential amplitude was within the normal range, indicating that the amount of acetylcholine per quantum and the post-synaptic response to acetylcholine are both normal. In 1982, Fukunaga et al. in freeze-fracture electron microscopy studies of neuromuscular junctions from LEMS patients described a significant reduction in presynaptic membrane active zones (AZs) [6]. This finding provided a morphological correlate of the electro-physiologic abnormalities, as the AZs are topographically related to the sites of synaptic exocytosis and are thought to represent the voltage-gated calcium channels (VGCCs) of the presynaptic membrane [6].

Lang and her colleagues in 1981 were the first to suggest an autoimmune basis for LEMS based on both clinical and experimental findings [7]. These included the clinical response to plasma exchange, which implicated a humoral factor, the improvement after immunosuppressive therapy, which also favoured an autoim-

Institute of Neurology, Catholic University, Largo F. Vito 1, 00168 Rome, Italy.
e-mail: a.evoli@vsb.it

mune pathogenesis, and the passive transfer of a presynaptic defect in neuro-muscular transmission to mice by the injection of LEMS IgG, suggesting the presence of antibodies to nerve terminal components.

One year later, a study by the two research groups led to a further demonstration of the autoimmune pathogenesis of the disease. In this case, using the passive transfer model of LEMS, they were able to reproduce both the electrophysiologic abnormalities and the presynaptic morphological lesions of the human disease. They found severe depletion of the AZs, which were aggregated into clusters [8].

Subsequent studies demonstrated that LEMS antibodies bind to AZs [9] and downregulate VGCCs [10] by antigenic modulation; that is, the divalent LEMS IgG antibodies cross-link adjacent AZ particles, the cross-linked particles aggregate into clusters and their density is reduced by accelerated internalisation [11].

The association of LEMS with SCLC was also clarified. It was found that SCLC cells express functional VGCCs and that IgG from LEMS patients reduces the number of calcium channels on cultured SCLC cell lines [12]. These data suggest that, in SCLC-associated LEMS, antibodies produced against VGCCs on cancer cells can cross-react with the same targets on the motor nerve terminal [1].

VGCCs and Presynaptic Proteins

VGCCs are pore-forming complexes of membrane proteins activated through membrane depolarisation; the resulting conformational changes permit selected ions to pass through the channel [13]. The calcium channels interact with many other proteins involved in the docking, priming and fusion of synaptic vesicles with the nerve terminal membrane [14].

VGCCs are multisubunit complexes formed by a central pore-forming α_1 subunit and several regulatory and/or auxiliary subunits, which include a cytoplasmic β subunit, a disulfide-linked $\alpha_2\delta$ subunit and, depending on the tissue of origin, a fifth subunit (the γ subunit in skeletal muscle) [13]. Auxiliary subunits do not carry Ca^{2+} currents themselves, but can modulate channel function through several mechanisms [15].

The α_1 is the largest of the VGCC subunits; it contains about 2000 amino acid residues organised in four repeated domains, each containing six transmembrane segments (S1-S6) [14] (Fig. 1). The α_1 subunit constitutes the voltage sensor, activates channel opening, binds different drugs and toxins and interacts with synaptic proteins. Several variants (A, B, C, D, E, F, G, H, I, S) have been cloned [13, 14].

VGCCs are distinguished on the basis of their electrophysiological and pharmacological properties, which, in turn, depend on the α_1 subunit [13, 14, 16]. Five classes of high-voltage activated (HVA) calcium channels have been identified (Table 1).

N-type and P/Q-type calcium channels are localised at high density in the presynaptic nerve terminals, and most evidence suggests that P/Q channels are primarily involved in acetylcholine release at the human neuromuscular junction [17, 18]. P-type channels are also found on cerebellar Purkinje cells, while Q-type

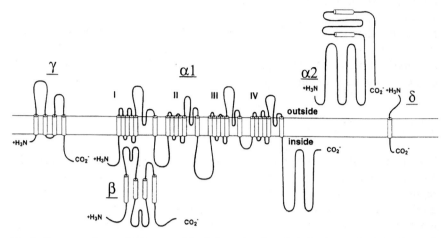

Fig. 1. Transmembrane folding model of the subunits of skeletal muscle Ca^{2+} channel. Predicted alpha elices are represented as cylindres. From [13], by permission

channels are associated with cerebellar granule cells and hippocampal neurons [13]. Pharmacologically, both channels can be blocked by ω-agatoxin IVA (P channels with high and Q channels with low affinity) and by ω-conotoxin MVIIC [19]. The α_{1A} subunits which form both P and Q channels are thought to derive from the same gene, the different sensitivities to ω-agatoxin IVA probably being due to the product of alternative mRNA splicing [13, 14].

Neurotransmitter release from synaptic vesicles requires the influx of Ca^{2+} through VGCCs, and a marked and rapid rise in the intracellular Ca^{2+} is required to trigger exocytosis. As the Ca^{2+} concentration rapidly declines by diffusion, threshold levels of intracellular Ca^{2+} are only reached in close proximity to the ion channels, which, in turn, are closely associated with exocytotic sites [20]. Central to the process of docking and fusion of the synaptic vesicles is the so-called SNARE complex formed by a set of proteins such as VAMP, SNAP-25 and syntaxin 1 [21, 22]. Synaptotagmin is a synaptic vesicle protein which binds Ca^{2+} and

Table 1. High-voltage-activated (HVA) calcium channels (Modified from [13, 14])

Channel type	Localisation in neurons	α_1 subunit	Blockers
L	Cell bodies, proximal dendrites	S, C, D ,F	Dihydropyridines
N	Nerve terminals, dendrites	B	ω-Conotoxin GVIA and MVIIA
P/Q	Nerve terminals, dendrites	A	ω-Agatoxin IVA and ω-Conotoxin MVIIC
R	Cell bodies, proximal dendrites	E	SNX-482

interacts with both SNARE proteins and calcium channels, functioning as a Ca^{2+} sensor for fast neurotransmitter release [20, 21]. It has been shown that P/Q and N calcium channels bind competitively with SNARE proteins (syntaxin 1 and SNAP-25) and synaptotagmin [14, 20, 21] via the synprint (synaptic protein interaction) site, which is localised to the cytoplasmic loop between domains II and III of the α_{1A} and α_{1B} subunits, respectively.

VGCCs Involved in LEMS

Earlier studies showed the presence of antibodies to ω-conotoxin GVIA-sensitive N-type calcium channels in serum from LEMS patients [23]. However, subsequent reports have established the higher frequency and pathogenic relevance of antibodies to ω-conotoxin MVIIC-sensitive P/Q channels [24-26]. This is in agreement with the fact that P/Q VGCCs are the principal channel subtype at the human neuromuscular junction (see above) [17], and down-regulation of these channels by LEMS antibodies impairs neuromuscular transmission. Indeed, serum antibodies that immunoprecipitate ^{125}I-ω-conotoxin MVIIC-labelled VGCCs are found in 85-90% of LEMS patients, whereas antibodies to ^{125}I-ω-conotoxin GIVA-labelled VGCCs are present in 40-50% of cases [24, 25].

Autonomic symptoms are a frequent complaint in LEMS. Their pathophysiology has been less extensively studied than that of muscle weakness, but a reduced neurotransmitter release due to impaired function of VGCCs can be hypothesised. Although VGCCs in autonomic neurons are both N- and P/Q type [27, 28], autonomic dysfunction in LEMS patients does not correlate with N-type channel antibodies [29]. In vitro studies have rather shown that LEMS IgG interfered with autonomic postganglionic transmission by reducing the number of P-type and Q-type channels [30-32].

Anti-P/Q-type VGCC antibodies were found positive in SCLC patients with cerebellar degeneration, in particular in those patients who were negative for anti-neuronal nuclear (anti-Hu) antibodies: this suggests that anti-VGCC antibodies can play a role in cerebellar dysfunction [33].

Antigen Specificity in LEMS

Pinto et al. tested the specificity of LEMS IgG for cloned human VGCCs and confirmed that the α_{1A} subunit, that is the pore-forming subunit of both P- and Q-type channels, is the target of these antibodies [34].

In the α_{1A} subunit molecule, the loop between the S5 and S6 segments (S5-S6 linker) in each domain is exposed extracellularly and is thus accessible to antibodies [14]. Serum antibodies to synthetic peptides corresponding to S5-S6 linkers of domains II and IV [35] and, more recently, to the recombinant protein derived from the DNA sequence coding for the same region of domain III [36], were found in LEMS patients. Moreover, immunisation with synthetic peptides corresponding to the S5-S6 linker of domain III induced an animal model of the disease [37].

Serum antibodies to the β subunit have been demonstrated in 23% of LEMS patients [38]. As the β subunit is intracellular, these IgG would not be pathogenic, but their production could be due to epitope spreading and their presence (like other antibodies to intracellular antigens), which would be detected by the radioimmunoassays that use solubilised VGCCs, could explain the lack of correlation between clinical severity and anti-VGCC antibody titre [38].

Possible Role of Antigens Other than VGCCs

Syntaxin, which is localised on the inner face of the plasma membrane, is not considered a relevant antigen in LEMS [39]. Synaptotagmin, being a synaptic vesicle protein, is exposed extracellularly during exocytosis. It has been proposed as an immunogen for LEMS on the basis of the induction of an animal model of the disease by immunisation with synthetic peptides corresponding to the N-terminus of rat synaptogamin 1 and the finding of serum antibodies to the recombinant protein in a proportion of LEMS patients [40]. Antibodies to ganglionic acetylcholine receptor, which could be responsible for autonomic dysfunction, have been found in some LEMS patients as well as in subjects with other paraneoplastic disorders [41].

Clinical Features

LEMS occurs more frequently in men. Cancer-associated disease is uncommon under the age of 40, while non-neoplastic LEMS, although rare in childhood, can present at any age [1, 42]. Most patients complain of weakness in the lower limb muscles, frequently associated with muscle stiffness and fatigability; ocular and oropharyngeal symptoms are usually mild and altogether less severe than in MG; threatening respiratory distress is uncommon. On examination, weakness is generally more pronounced in the proximal muscles of the leg and, less frequently, of the arm. Mild ptosis is the most frequent ocular sign; although diplopia is a common complaint, ophthalmoplegia is generally not observed [1]. Muscle strength increases immediately after exercise (facilitation), then decreases if muscular effort is sustained; this finding is quite characteristic, but it is not always easy to detect [42]. Tendon reflexes are hypoactive or absent in most cases but can be elicited after brief maximum contraction of the tested muscles.

Autonomic symptoms are common, being reported in up to 80% of patients [1]; the most common complaint is dryness of the mouth, followed in order of prevalence by sexual impotence, constipation, postural hypotension, increase or decrease in sweating and blurred vision [1, 29, 42]. Autonomic tests show abnormal results in most patients, even in the absence of clinical symptoms and signs, showing both parasympathetic and sympathetic dysfunction [29].

In most cases, the disease onset is gradual and the course is slowly progressive; however, weakness can begin acutely, especially in cancer patients, and respiratory failure can be an early sign in rare instances [1].

Malignancy is associated in nearly 50% of patients and is represented by SCLC in 80-90% of cases [1, 43]; conversely LEMS occurs in only 3% of patients with SCLC. LEMS symptoms generally precede the diagnosis of the tumour. The risk of an underlying SCLC at LEMS presentation is estimated at nearly 60%; it declines markedly after 2 years and is low after 4 years [1]. In patients with SCLC, LEMS can coexist with other paraneoplastic neurological syndromes, especially cerebellar degeneration [33]. An association with other tumours, such as lymphosarcoma, thymoma or carcinomas of the breast, stomach, colon, bladder or kidney has also been reported [42]. Non-neoplastic LEMS can be associated with autoimmune disorders such as thyrotoxicosis, vitiligo, rheumatoid arthritis, MG and pernicious anaemia [1].

Diagnosis

Clinical features, namely proximal muscle weakness and hypoactive tendon reflexes with facilitation, especially when associated with autonomic dysfunction, are strongly suggestive of LEMS. Edrophonium (Tensilon) injection can induce some improvement in strength, but its effect is generally not so dramatic as in MG [42].

Electrodiagnosis

Electromyography (EMG) is diagnostic, provided that the classical triad is demonstrated: (1) an abnormally small amplitude of the resting compound muscle action potential (CMAP), (2) a further decrement of CMAP with low-rate repetitive nerve stimulation (RNS) and (3) a CMAP increment (facilitation) during high-rate RNS or immediately after maximal voluntary contraction (MVC). Although, in LEMS, weakness is more prominent in proximal muscles, EMG findings are more easily detected in distal muscles [44].

A low CMAP amplitude at rest (ranging from 0.1 to 6mV) is seen in 95% of patients [45]. This finding is considered the best electrophysiological index of LEMS severity [46], but it can be absent in the early stages of the disease and is not specific as it can be observed in nerve and muscle diseases [42].

The demonstration of facilitation represents an essential step in the electrophysiological diagnosis of LEMS. Figure 2 shows a significant facilitation during RNS at 50 Hz in two patients with LEMS. As high-rate RNS is painful for the patient, facilitation is preferentially evoked as post-exercise facilitation, by delivering a supramaximal nerve stimulus immediately (within 5 s) after 10-15 s MVC of the tested muscle [42, 44]. When the increase in CMAP amplitude is greater than 100%, it is considered as diagnostic of LEMS: when greater than 50% it is suggestive of LEMS, but such an increase can also be seen in MG [42]. False-neg-

50 Hz

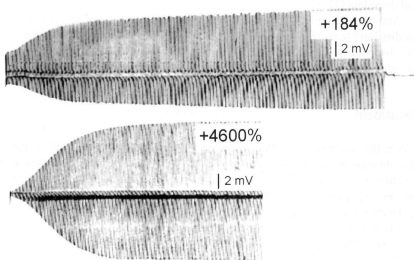

Fig. 2. Repetitive nerve stimulation (RNS) in the hypothenar muscles of two patients with Lambert-Eaton myasthenic syndrome. Marked increase in compound muscle action potential amplitude (facilitation) during high-rate (50 Hz) RNS

ative results can occur both in cases of lack of co-operation on the part of the patient and in cases of severe long-standing weakness [1, 47]. To increase the test sensitivity in severely affected patients, the application of low-rate RNS during voluntary contraction has been proposed [48].

A decrement of CMAP greater than 10% with low-rate (3-5 Hz) RNS is seen in nearly all LEMS patients; this finding is by no means specific as it is found in MG and in some neurogenic conditions [45]. Single-fibre EMG shows increased jitter and blocking; jitter abnormality, typically, decreases with higher discharge rate [45]. However, this pattern is not always found in LEMS patients, and it is not diagnostic as it can also be seen in MG [42].

Serological Diagnosis

Serum antibodies against P/Q type VGCCs can be measured by a radioimmunoassay using ^{125}I-ω-conotoxin MVIIC-labelled cerebellum extract as antigen. This assay gives positive results in 82-95% of LEMS patients [24, 31, 40, 49, 50]. Antibody levels do not correlate with EMG abnormalities between different patients, but in longitudinal studies on individual patients, EMG improvement was found to be associated with a decline in anti-VGCC titre [49].

While Lennon et al. reported a high frequency of anti-P/Q-type VGCC antibodies in diseases other than LEMS [24], in subsequent studies positive titres were found in isolated cases of amyotrophic lateral sclerosis, in a few subjects with

SCLC and in 12-15% of patients with SCLC-associated paraneoplastic neurological disorders [31, 50, 51].

Anti-Hu antibodies and, more recently, anti-Purkinje cell cytoplasmic antibodies have been reported in LEMS as well as in other paraneoplastic neurological disorders [33, 52]. As these IgG are serological markers of neurological diseases related to SCLC, their detection could favour an early cancer diagnosis.

Treatment

Upon the diagnosis of LEMS, a search for malignancy is imperative. Patients should undergo computed tomography (CT) of the chest and, if they are at risk for lung cancer, bronchoscopy should be done even if radiological examinations are normal [42]. If the initial evaluation is negative, tumour surveillance with periodic CT examinations should be continued for at least 5 years after the onset of the disease.

In paraneoplastic LEMS, aggressive treatment of the tumour is always justified as it frequently leads to clinical improvement [53]. In non-cancer cases, treatment depends on symptom severity: patients with mild disease can be given symptomatic drugs only, whereas severely affected patients often need immunosuppressive therapy. When required, LEMS treatment is also justified in cancer-associated disease, although the possible negative effects of immunosuppression in patients with malignancy must be taken into consideration.

Symptomatic treatment with 3,4-diaminopyridine (DAP) proved effective in most LEMS patients, relieving both muscle and autonomic symptoms [54]. Acting as a voltage-gated potassium channel blocker, this drug prolongs the nerve action potential, so lengthening the open time of VGCCs and increasing the release of acetylcholine from the nerve terminal [55]. DAP is administered at starting doses of 5-10 mg three to five times a day; if needed, the dose can be gradually increased up to a maximum of 100 mg a day [43, 56]. Side effects are usually mild (transient paraesthesias), but seizures can occur in patients taking daily doses greater than 100 mg, and even at lower dosages in patients with cerebral lesions [42, 56].

Guanidine hydrochloride acts by increasing the intracellular calcium concentration, which ultimately leads to an increased output of acetylcholine from the nerve terminal [42]. It should be considered as a second-choice agent in the symptomatic treatment of LEMS as, although it proved effective, serious side effects such as bone marrow suppression, renal failure, cardiac arrhythmia and hepatic toxicity can occur at doses greater than 1000 mg a day [42, 56, 57].

Pyridostigmine as a cholinesterase inhibitor can induce some improvement in LEMS patients and should be tried, at doses ranging from 30 to 60 mg three to four times a day, both as a single drug and in association with other symptomatic agents, whose therapeutic effect it can enhance [42, 56, 57].

In patients with disabling weakness, either with or without tumour, immunosuppressive therapy should be considered. Long-term pharmacological therapy with corticosteroids and/or azathioprine is given according to the same guidelines as in MG. Prednisone/prednisolone is usually given at high daily doses at the start of therapy, then on alternate day administration, slowly tapering drug dosage to the minimum effective dose. Azathioprine can be used as a single drug in patients with contraindications to corticosteroids, or together with prednisone for its steroid sparing effect. Corticosteroid therapy is not contraindicated by the presence of cancer [56].

Short-term treatments such as plasma exchange (PE) and intravenous immunoglobulin (IVIG) have both proved effective in LEMS; however, they induce an improvement which is temporary unless pharmacological immuno-suppression is given in combination [43, 56]. The effect of PE shows a time-to-onset longer than that observed in MG (10 days compared to 2 days), probably reflecting the fact that VGCCs have a longer turnover time than acetylcholine receptors [58].

In conclusion, being an antibody-mediated disease, LEMS benefits from immunosuppressive treatment. However, symptomatic therapy with DAP plus pyri-dostigmine (or guanidine plus pyridostigmine) should be tried in all LEMS patients, with or without cancer. The transient improvement induced by PE or IVIG may be of benefit in cancer patients, who are likely to improve with anti-neoplastic therapy. Otherwise, short-term treatments are used in patients with severe weakness in association with pharmacological immunosuppression. Long-term treatment with corticosteroids with or without azathioprine is indicated in all patients with disabling disease who do not respond satisfactorily to symptomatic drugs.

References

1. O'Neill JH, Murray NMF, Newsom-Davis J (1988) The Lambert-Eaton myasthenic syndrome. A review of 50 cases. Brain 111:577-596
2. Anderson HJ, Churchill-Davidson HC, Richardson AT (1953) Bronchial neoplasm with myasthenia: prolonged apnoea after administration of succinylcholine. Lancet ii:1291-1293
3. Eaton LM, Lambert EH (1957) Electromyography and electrical stimulation in diseases of the motor unit: observations on a myasthenic syndrome associated with malignant tumors. JAMA 163:1117-1124
4. Lambert EH, Rooke ED, Eaton LM, Hodgson CH (1962) Myasthenic syndrome occasionally associated with bronchial neoplasm: neurophysiologic studies. In: Viets HR (ed) Myasthenia gravis. Thomas, Springfield, Ill, pp 362-410
5. Lambert EH, Elmquist D (1971) Quantal components of end-plate potentials in the myasthenic syndrome. Ann N Y Acad Sci 183:183-199
6. Fukunaga H, Engel AG, Osame M, Lambert EH (1982) Paucity and disorganization of presynaptic membrane active zones in the Lambert-Eaton myasthenic syndrome. Muscle Nerve 5:686-697
7. Lang B, Newsom-Davis J, Wray D et al (1981) Autoimmune aetiology for myasthenic (Eaton-Lambert) syndrome. Lancet ii:224-226
8. Fukunaga H, Engel AG, Lang B et al (1983) Passive transfer of Lambert-Eaton myasthenic syndrome with IgG from man to mouse depletes the presynaptic membrane active zones. Proc Natl Acad Sci USA 80:7636-7640

9. Fukuoka T, Engel AG, Lang B et al (1987) Lambert-Eaton myasthenic syndrome: I. Early morphological effects of IgG on the presynaptic membrane active zones. Ann Neurol 22:193-199

10. Lang B, Newsom-Davis J, Peers C et al (1987) The effects of myasthenic syndrome antibody on presynaptic calcium channels in the mouse. J Physiol 390:257-270

11. Nagel A, Engel AG, Lang B et al (1988) Lambert-Eaton myasthenic syndrome IgG depletes presynaptic membrane active zones particles by antigenic modulation. Ann Neurol 24:552-558

12. Roberts A, Perera S, Lang B et al (1985) Paraneoplastic myasthenic syndrome IgG inhibits $^{45}Ca^{2+}$ flux in a human small cell carcinoma line. Nature 317:737-739

13. Greenberg DA (1999) Neuromuscular disease and calcium channels. Muscle Nerve 22:1341-1349

14. Catterall WA (1998) Structure and function of neuronal Ca^{2+} channels and their role in neurotransmitter release. Cell Calcium 24:307-324

15. Walker D, De Waard M (1998) Subunit interaction sites in voltage-dependent Ca^{2+} channels: role in channel function. Trends Neurosci 21:148-154

16. Birnbaumer L, Campbell KP, Catterall WA et al (1994) The naming of voltage-gated calcium channels. Neuron 13:505-506

17. Protti DA, Reisin R, Macklinley TA, Uchitel OD (1996) Calcium channel blockers and transmitter release at the normal human neuromuscular junction. Neurology 46:1391-1396

18. Day NC, Wood SJ, Ince PG et al (1997) Different localization of voltage-dependent calcium channel α_1 subunits at the human and rat neuromuscular junction. J Neurosci 17:6226-6235

19. Liu H, De Waard M, Scott VES et al (1996) Identification of three subunits of the high affinity ω-conotoxin MVIIC-sensitive Ca^{2+} channel. J Biol Chem 271:13801-13810

20. Stanley EF (1997) The calcium channel and the organization of the presynaptic transmitter release face. Trends Neurosci 20:404-409

21. Turner KM, Burgoyne RD, Morgan A (1999) Protein phosphorylation and the regulation of synaptic membrane traffic. Trends Neurosci 22:459-464

22. Chen YA, Scales SJ, Patel SM et al (1999) SNARE complex formation is triggered by Ca^{2+} and drives membrane fusion. Cell 97:165-174

23. Sher E, Gotti C, Canal N et al (1989) Specificity of calcium channel autoantibodies in Lambert-Eaton myasthenic sindrome. Lancet ii:640-643

24. Lennon VA, Kryzen TJ, Griesman GE et al (1995) Calcium-channel antibodies in the Lambert-Eaton syndrome and other paraneoplastic syndromes. N Engl J Med 332:1467-1474

25. Motomura M, Johnston I, Lang B et al (1995) An improved diagnostic assay for Lambert-Eaton myasthenic syndrome. J Neurol Neurosurg Psychiatry 58:85-87

26. Satoh Y, Hirashima N, Tokumaru H et al (1998) Lambert-Eaton syndrome antibodies inhibit acetylcholine release and P/Q-type Ca^{2+} channels in electric ray nerve endings. J Physiol 508:427-438

27. Waterman SA (1996) Multiple subtypes of voltage-gated calcium channel mediate transmitter release from parasympathetic neurons in the mouse bladder. J Neurosci 16:4155-4161

28. Waterman SA (1997) Role of N-, P-, and Q-type voltage-gated calcium channels in transmitter release from sympathetic neurons in the mouse vas deferens. Br J Pharmacol 120:393-398

29. O'Sulleabhain P, Low PA, Lennon VA (1998) Autonomic dysfunction in the Lambert-Eaton myasthenic syndrome. Serologic and clinical correlates. Neurology 50:80-93

30. Waterman S, Lang B, Newsom-Davis J (1997) Effect of Lambert-Eaton myasthenic syndrome antibodies on autonomic neurons in the mouse. Ann Neurol 42:147-156

31. Lang B, Waterman S, Pinto A et al (1998) The role of autoantibodies in Lambert-Eaton myasthenic syndrome. Ann N Y Acad Sci 841:596-605

32. Houzen H, Hattori Y, Kanno M et al (1998) Functional evaluation of inhibition of autonomic transmitter release by autoantibody from Lambert-Eaton myasthenic syndrome. Ann Neurol 43:677-680

33. Mason WP, Graus F, Lang B et al (1997) Small-cell lung cancer, paraneoplastic cerebellar degeneration and the Lambert-Eaton myasthenic syndrome. Brain 120:1279-1300

34. Pinto A, Gillard S, Moss F et al (1998) Human antibodies specific for the α_{1A} calcium channel subunit reduce both P-type and Q-type calcium currents in cerebellar neurons. Proc Natl Acad Sci USA 95:8328-8333

35. Takamori M, Iwasa K, Komai K (1998) Antigenic sites of the voltage-gated calcium channels in the Lambert-Eaton myasthenic syndrome. Ann N Y Acad Sci 841:625-635

36. Iwasa K, Takamori M, Komai K, Mori Y (2000) Recombinant calcium channel is recognized by Lambert-Eaton myasthenic syndrome antibodies. Neurology 54:757-759

37. Komai K, Iwasa K, Takamori M (1999) Calcium channel peptide can cause an autoimmune-mediated model of Lambert-Eaton myasthenic syndrome in rats. J Neurol Sci 166:126-130

38. Verschuuren JJ, Dalmau J, Tunnel R et al (1998) Antibodies against the calcium channel β-subunit in the Lambert-Eaton myasthenic syndrome. Neurology 50:475-479

39. Hajela RK, Atchison WD (1995) The proteins synaptotagmin and syntaxin are not general targets of Lambert-Eaton myasthenic syndrome autoantibodies. J Neurochem 64:1245-1251

40. Takamori M, Maruta T, Komai K (2000) Lambert-Eaton myasthenic syndrome as an autoimmune calcium-channelopathy. Neurosci Res 36:183-191

41. Vernino S, Adamski J, Kryzer TJ et al (1998) Neuronal nicotinic ACh receptor antibody in subacute autonomic neuropathy and cancer related syndromes. Neurology 50:1806-1813

42. Sanders DB (1995) Lambert-Eaton myasthenic syndrome: clinical diagnosis, immune-mediated mechanisms, and update on therapies. Ann Neurol 37 (S1):S63-S73

43. Lennon VA, Lambert EH, Whittingham S, Fairbanks V (1982) Autoimmunity in the Lambert-Eaton syndrome. Muscle Nerve 5:S21-S25

44. Maddison P, Newsom-Davis J, Mills KR (1998) Distribution of electrophysiological abnormality in Lambert-Eaton myasthenic syndrome. J Neurol Neurosurg Psychiatry 65:213-217

45. Oh SJ (1988) Electromyography. Neuromuscular transmission studies. Williams and Wilkins, Baltimore

46. Oh SJ, Kim DE, Kuruoglu R et al (1996) Electrophysiological and clinical correlations in the Lambert-Eaton myasthenic syndrome. Muscle Nerve 19:903-906

47. Oh SJ (1989) Diverse electrophysiological spectrum of the Lambert-Eaton myasthenic syndrome. Muscle Nerve 12:464-469

48. Lo Monaco M, Milone M, Padua L, Tonali P (1997) Combined low-rate nerve stimulation and maximal voluntary contraction in the detection of compound muscle action potential facilitation in Lambert-Eaton myasthenic syndrome. Muscle Nerve 20:1207-1208

49. Motomura M, Lang B, Johnston I et al (1997) Incidence of serum anti-P/Q type and N-type antibodies in the Lambert-Eaton myasthenic syndrome. J Neurol Sci 147:35-42

50. Nakao YN, Motomura M, Suenaga A et al (1999) Specificity of ω-conotoxin MVIIC-binding and blocking calcium channel antibodies in Lambert-Eaton myasthenic syndrome. J Neurol 246:38-44

51. Voltz , Carpentier AF, Rosenfeld MR et al (1999) P/Q type voltage-gated calcium channel antibodies in paraneoplastic disorders of the central nervous system. Muscle Nerve 22:119-122

52. Vernino S, Lennon VA (2000) New Purkinje cell -voltage-gatedbody (PCA-2): marker of lung cancer-related neurological autoimmunity. Ann Neurol 47:297-305

53. Chalk CH, Murray NMF, Newsom-Davis J et al (1990) Response of the Lambert-Eaton myasthenic syndrome to the treatment of the associated small-cell lung carcinoma. Neurology 40:1552-1556

54. Sanders DB, Massey JM, Sanders LL, Edwards LJ (2000) A randomized trial of 3,4-diaminopyridine in Lambert-Eaton myasthenic syndrome. Neurology 54:603-607

55. Thomsen RH, Wilson DF (1983) Effects of 4-aminopyridine and 3,4-diaminopyridine on transmitter release at the neuromuscular junction. J Pharmacol Exp Ther 227:260-265

56. Newsom-Davis J (1998) A treatment algorithm for Lambert-Eaton myasthenic syndrome. Ann N Y Acad Sci 841:817-822

57. Oh SJ, Kim DS, Kwon KH et al (1998) Wide spectrum of symptomatic treatment in Lambert-Eaton myasthenic syndrome. Ann N Y Acad Sci 841:827- 831

58. Newsom-Davis J, Murray JNM (1984) Plasma exchange and immunosuppressive drug treatment in the Lambert-Eaton myasthenic syndrome. Neurology 34:480-485

Chapter 4

Immunotherapy of Myasthenia Gravis

C. Antozzi[1], F. Baggi[1], F. Andreetta[1], M. Milani[1], A. Annoni[1], P. Bernasconi[1], R. Mantegazza[1], F. Cornelio[2]

Myasthenia gravis (MG) is an acquired autoimmune disease of the neuromuscular junction mediated by antibodies against the nicotinic acetylcholine receptor (anti-AChR Ab). Anti-AChR Ab, by cross-linking AChR molecules and activating the complement cascade, cause destruction of junctional AChR and simplification of the post-synaptic membrane [1, 2]. The reduced number of functional AChR accounts for the impaired neuromuscular transmission, leading to the muscle weakness and fatigability typical of MG patients. For a review of the clinical features of MG, see [3].

The natural course of MG, as reported in a large series of patients, has been considerably modified by the improvement of intensive care technology and by the introduction of immunosuppressive and immunomodulating treatments [4, 5]. This review will focus on the different therapeutic options available for MG, including thymectomy, immunosuppressive (corticosteroids and immunosuppressants) and immunomodulating treatments (plasmapheresis, immunoadsorption and intravenous immunoglobulins). We will also briefly consider recent antigen-specific immunomodulating approaches investigated in experimental autoimmune myasthenia gravis (EAMG), the animal model of MG.

Thymectomy

Since the first report by Blalock [6], thymectomy has been used as a measure to modify the course of MG, based on the assumption that sensitisation to the AChR, and possibly maintenance of the autoimmune attack, might occur in the thymus [1]. To date, the clinical outcome of MG after thymectomy has not been evaluated in a controlled fashion; moreover, due to the variable course of the disease, differences in surgical techniques and ongoing long-term immunosuppression, the role of thymectomy in improving MG is difficult to establish. Nevertheless, the majority of reported series suggest that the outcome of MG patients is improved by thymectomy. The recommended approach is through the extended trans-sternal

[1] Myopathology and Immunology Unit, Department of Neuromuscular Diseases; [2] Department of Neuromuscular Diseases, National Neurological Institute "Carlo Besta", Via Celoria 11, 20133 Milan, Italy. e-mail: car.antozzi@xquasar.com

technique, which provides wide visualisation of the mediastinal space, enabling removal of the thymus and ectopic thymic tissue in the mediastinal fat that extends from the thyroid gland to the pericardium [7]. In this respect, the transcervical approach should be discouraged as it may result in incomplete removal. Indications for surgery include patients with generalised MG, but no agreement exists regarding the minimum and maximum age at which patients should be operated on. Thymectomy is generally not performed in small children, or in patients with ocular MG.

The safety of the procedure has considerably improved and the mortality rate is less than 1%. The most common morbidity is respiratory failure from MG crisis in the immediate postoperative period. The occurrence of respiratory insufficiency can be reduced considerably by stabilising the patient's condition before surgery with adequate immunosuppression, and plasmapheresis if needed. Thymectomy should never be performed as an "emergency procedure" in patients showing signs of severe generalised or bulbar MG. Data from uncontrolled series suggest that the best response is obtained in young female patients operated on early in the course of the disease; the thymus is usually hyperplastic in these patients.

A new approach suitable for extended thymectomy that does not require a median sternotomy has recently been introduced [8, 9]. The video-assisted thoracoscopic extended thymectomy (VATET) allows a several-fold magnified observation of the mediastinal space for maximum removal of the thymus and surrounding fat. This technique can be performed with either a mono- or bilateral approach, is well tolerated, minimises side effects, shortens hospitalisation and is better accepted by patients.

The impact of thymectomy on MG has been reported in an evidence-based review of the literature by the Quality Standards Subcommittee of the American Academy of Neurology [10]. The authors concluded that there is a positive association between thymectomy and improved MG outcome, but this association becomes conflicting whenever multiple confounding variables are taken into account. The authors underline the shortcomings of reported series and the need for controlled studies, and recommend thymectomy "as an *option* to increase the probability of remission or improvement" [10].

Cortisteroids and Immunosuppressive Drugs

Corticosteroid therapy is rapidly effective in the majority of MG patients and remains the treatment of choice for the severe forms of the disease that do not respond to anticholinesterase therapy. The introduction of corticosteroids has considerably improved the management of patients with bulbar MG, and their maximum efficacy is usually evident within 2 months from the beginning of treatment [11]. Several schedules have been reported. A widely used regimen begins with the daily administration of prednisone 1 mg/kg/day; once an adequate response is obtained, treatment can be shifted to an alternate-day regimen, and then slowly tapered to the minimum effective dose to minimise side effects.

For safety reasons, treatment with high-dose prednisone should always be started in hospitalised patients because of the possible exacerbation of myasthenic symptoms during the first week of treatment. Intravenous pulse methylprednisolone has been also used with positive results in patients with MG exacerbations, followed by maintenance treatment with prednisone [12].

Clinical experience over more than three decades has confirmed the efficacy of corticosteroids in patients with severe MG. However, it must be emphasised that MG is a chronic disease requiring long-term drug regimens, the side effects of which must not be overlooked. The association with other drugs must be taken into consideration so as to maximise immunosuppression while at the same time trying to reduce, and possibly withdraw, corticosteroids. The efficacy of prednisone given alone or in combination with azathioprine, evaluated retrospectively in a large sample of MG patients, showed that about 75% of patients had a positive outcome (including remission and asymptomatic and improved cases) [13].

Azathioprine is the most widely used immunosuppressive and steroid-sparing agent in MG [14-17]. Its efficacy and steroid-sparing effect have been demonstrated by a controlled trial comparing prednisolone alone versus prednisolone combined with azathioprine [18]. It was shown that patients treated with the combination of the two drugs experienced a better outcome, longer remissions and fewer side effects. The recommended dose for azathioprine is 2-3 mg/kg per day, starting with 50 mg/day and increasing the dose by 50 mg weekly until the full dose is achieved. The onset of clinical improvement is usually delayed for several months, and it is advisable to continue treatment with azathioprine for at least 1 year before considering the drug ineffective. Azathioprine is well tolerated by the majority of patients and response rates up to 70% have been reported. Blood tests for haematological side effects (bone marrow depression) and hepatotoxicity should be performed periodically for as long as the patient is receiving the treatment. Mild to moderate gastric discomfort is frequently experienced at the beginning of treatment, and in a minority of patient requires withdrawal of drug. There is no general consensus on how long treatment with azathioprine should be continued; reactivation of the disease has been reported after discontinuation of long-term treatment [19].

Cyclosporine can be an alternative immunosuppressant in patients who do not respond to or do not tolerate azathioprine [20, 21]. Cyclosporine has been evaluated in a 12-month double-blind placebo-controlled trial with the conclusion that it can be effective in some MG patients [22]. Patients must be monitored carefully for signs of nephrotoxicity and hypertension, particularly when the daily dose exceeds 6 mg/kg perday. In the same way as for azathioprine, the onset of improvement can occur several months after the beginning of treatment.

The experience with *cyclophosphamide* is limited, mainly because of its toxicity, particularly myelosuppression, haemorrhagic cystitis, nausea, alopecia and increased risk of malignancies. The occurrence of sterility represents a further major limitation to its use. Nevertheless, cyclophosphamide can be effective in patients with severe, refractory forms of MG after failure with other combined therapies [23, 24]. The daily dose is similar to that of azathioprine (2-3 mg/kg per day). Hydration should be increased to reduce accumulation of toxic metabolites in the bladder, as well as the incidence of hemorrhagic cystitis. Alternatively, cyclophosphamide can be administered by periodic intravenous infusions (the average dose is 0.5-1g/m^2 body surface); the intravenous infusion requires hydration, bladder protection and anti-emetic therapy. Blood cell counts must be frequently monitored whatever route of administration is used.

Immunomodulating Therapies

Treatment schedules using prednisone and/or other immunosuppressive agents, though effective, may take time to exert their clinical efficacy. Moreover, patients with the most severe forms of the disease (i.e. bulbar myasthenia) are at risk of respiratory insufficiency. Immunomodulating therapies can be of great help in these patients as well as in those with myasthenic crisis to obtain a rapid improvement of their clinical status.

Since its first application in 1976 [25], *plasma exchange* (PE) has been the treatment of choice for MG crises and has considerably modified the natural course of the disease. The efficacy of PE has never been evaluated in a controlled fashion, but the striking clinical results and time-related association between plasma removal and improvement are proof of its efficacy. PE is usually considered a short-term treatment, the indications being MG crisis, severe deterioration, worsening related to the start of corticosteroid therapy, and, in selected patients, preparation for thymectomy. Improvement is usually very rapid and can be observed within a few days. The clinical effect lasts for only a few weeks unless the treatment is combined with effective immunosuppression. The recommended protocols exchange one plasma volume every other day, for a mean of two or three sessions. In our experience, two exchanges of one plasma volume each are effective in 70% of MG patients with bulbar involvement [26]. There is no clear evidence that intensive protocols are needed if a few sessions do not exert a significant clinical effect.

The use of PE as a chronic treatment has not been investigated in detail, but periodical removal of plasma can be of help in patients showing a very slow or inadequate response to ongoing immunosuppression. The periodic removal of IgG, and hence anti-AChR Ab, by *staphylococcal protein A immunoadsorption* is dramatically effective in patients with severe, treatment-resistant MG. The procedure removes IgG selectively from the patient's plasma and does not require any replacement fluid. Moreover, the on-line regeneration of protein A after each cycle of immunoadsorption allows the treatment of unlimited amounts of plasma. This technique is indicated in selected patients who fail to respond to prolonged immunosuppression or require frequent PE to maintain satisfactory improvement. Interestingly, we obtained a considerable improvement in MG patients who did not respond to PE or high-dose intravenous immunoglobulins [27, 28]. The efficacy of immunoadsorption is probably due to a considerably higher IgG-removing capacity compared to conventional PE.

High dose intravenous immunoglobulins (IVIG) in MG are an alternative to PE and share the same indications [29]. Several uncontrolled series have been reported in the literature, and the most widely used treatment protocol consists of five consecutive daily infusions of 400 mg of immunoglobulins per kg/day. The majority of patients improve after IVIG, the degree of improvement is variable and the clinical response does not seem to be as rapid as that observed with PE. The important issue of the comparison between IVIG and PE was addressed in a randomised trial involving patients with acute forms of the disease [30]. The primary end point of the study was the change in clinical score recorded before randomisation and at day 15. No significant differences were found between the two treatments. The effect of IVIG and PE has been recently evaluated retrospectively in patients with MG crisis [31], concluding that PE was associated with a better ventilatory status at two weeks and clinical outcome at one month. A positive response to PE after failure with IVIG has been reported in a small series of patients [32]. Apart from possible differences in rapidity of action, IVIG are an important treatment option in patients with severe MG when PE is not readily available, or not feasible because of inadequate vascular access or contraindications to extracorporeal circulation.

Conclusions Regarding Conventional Immunotherapies

The experience with conventional immunosuppressive treatment in a large series of patients demonstrated that a considerable proportion of MG patients improved with pharmacological treatment [13]. However, the percentage of asymptomatic and improved patients remained stable over a 9-year follow-up, and, most importantly, the clinical outcome was not significantly modified in about 25% of cases. Clinical experience demonstrated also that long-term immunosuppression is needed to obtain stable control of the disease. In this connection, it must be emphasised that medical treatment of MG can be a source of severe side effects, partly due to the drugs themselves, and partly because of dosage and long-term administration (particularly corticosteroids). These drawbacks suggest the need for more specific and less toxic immunotherapeutic approaches. The ideal treatment should inhibit the immune response against the AChR specifically and permanently, without interfering with the general functioning of the immune system.

Information available on the antigenic structure and epitope mapping of the AChR allows the design of new therapeutic approaches able to downregulate the autoimmune response. Antigen-specific immunotherapies are not yet available for MG. However, a considerable amount of data have emerged from recent work on EAMG [33], which provides a suitable animal model for the investigation of different specific immunotherapies in vivo.

Immunotherapy of EAMG

EAMG can be induced in rabbits, guinea pigs, mice and rats by immunisation with AChR purified from *Torpedo californica* (TAChR). The immunological features of EAMG have been extensively investigated in C57/Bl6 (B6) mice and Lewis rats. In both models, as well as in MG, pathogenic anti-AChR Ab react predominantly with the α-subunit of the AChR [34, 35]. The complement-fixing $IgG_{2a,b}$ isotypes of anti-AChR Ab are thought to be crucial in the pathogenesis of the experimental disease [36]. Although the triggering events are unknown, AChR are processed and presented as immunodominant peptides to T helper cells in the context of specific MHC class II molecules. The role of these molecules has been demonstrated by studies performed in bm12 mice that carry a three-amino-acid substitution in the β chain of the I-A molecule making the strain resistant to EAMG [37]. Moreover, MHC class II knock-out mice are also resistant to induction of EAMG [38]. In B6 mice, the immunodominant T cell epitopes encompass the AChR α-subunit sequences α111-126, α146-162, α182-198, and α360-368 [39, 40]; the sequence α100-116 is immunodominant in Lewis rats [41].

Several strategies potentially able to interfere at different levels with the autoimmune reaction have been investigated in EAMG, a T-cell-dependent, B-cell-mediated disease. Complex approaches have considered the inhibition of activated T cells with conjugates of *Torpedo* AChR attached to toxins, interference of costimulation by means of CTLA4Ig, depletion of AChR-specific B cells, and the use of modified antigen presenting cells (APC) to avoid activation of antigen-specific T cells. All of these approaches have yielded promising results in vitro that prompted their investigation in the animal model [42-45]. However, the most encouraging data have emerged from studies on the effect of the systemic or mucosal administration of the native antigen, recombinant fragments or peptides.

Tolerance induced by the oral administration of antigens, which results in suppression of the immune response to the fed protein, has been investigated in several models of autoimmune disease [46]. The mechanisms underlying the protective effect are related, at least in part, to the amount and type (native vs. recombinant or peptides) of antigen given. Low doses induce active suppression mediated by release of Th2/3 cytokines, while high doses cause anergy of specific T lymphocytes. The administration of antigens through the nasal route, though less investigated, seems to share similar mechanisms.

Nasal and *oral tolerisation* of EAMG were first obtained by the administration of TAChR in Lewis rats, with a significant degree of protection [47-49]. However, when TAChR was given orally after immunisation, to inhibit an ongoing autoimmune response to the AChR, the clinical effect was limited and an increased production of anti-AChR Ab was found, suggesting that the native receptor, which is known to be highly immunogenic [50], had stimulated the antibody response. *Recombinant fragments of the AChR* or peptides corresponding to *immunodominant T cell epitopes* represent an alternative to the administration of the native antigen. Indeed, EAMG has been tolerised by the nasal administration route [51], and the ongoing disease has been suppressed by oral treatment with a fragment corresponding to the extracellular domain of the human AChR α-subunit (Hα1-205) [52]. Administration of the recombinant fragment induced a shift from Th1- to a Th2/Th3-type T cell response as shown by cytokine production and changes in anti-AChR Ab isotypes. Interestingly, the degree of refolding of the recombinant protein has been investigated and found to have an important role in determining the protective effect. The native conformation of recombinant fragments of the AChR is potentially immunogenic and not suitable for treatment purposes, compared to the less native or denatured recombinant antigens [53]. In this regard, specific immunosuppression of EAMG had already been demonstrated by immunisation with a denatured preparation of the TAChR in rabbits [54].

Alternatively, *synthetic peptides* corresponding to immunodominant T cell epitopes have been investigated, particularly in the mouse model. The parenteral administration of the epitope α146-162 or of a pool of immunodominant epitopes prevented EAMG in B6 mice [55-56]. Similar findings were also found after nasal delivery of the same peptide sequences [57]. The disease has also been prevented in B6 mice by oral feeding with high doses of the epitope α146-162 [58]. As observed with the high-dose, subcutaneous administration, oral feeding with peptide α146-162 induced tolerance and suppressed T cell responses to the peptide itself, to the whole native antigen (TAChR) and to a subdominant epitope (α182-198) that was not used for treatment. This suggests a spreading of tolerance to other T cell epitopes [55, 58]. The protective effect induced by feeding high doses of the peptide was associated with downregulation of Th1/Th2 cytokines; no evidence of production of TGF-β was found in Peyer's patches [58].

Two single substituted human AChR α-subunit peptides, Hα195-212 and Hα259-271, as well as a dual analog of these two immunodominant T cell epitopes in MG, and in SJL and BALB/c mice, have been investigated for their capacity to interfere with specific autoimmunity in vitro and prevent the induction of EAMG in vivo [59-61]. The effect of these altered peptide ligands on the activity of tyrosine kinase and phospholipase C was studied in order to explain their inhibitory effect on antigen-specific T cell responses. The altered peptides were shown to interfere with the signal transduction pathway by inhibiting phospholipase C activity induced by myasthenogenic T cell epitopes [62]. Moreover, the oral administration of the same dual analog was effective on ongoing EAMG induced in B6 mice [63].

The experience with peptides in the Lewis rat model is limited compared to that with EAMG in B6 mice. However, tolerance has been induced by a soluble complex of MHC class II and peptide α100-116, which is immunodominant in Lewis rats [64]. In this model, the MHC:peptide complexes, probably owing to absence of a second co-stimulatory signal, were able to suppress antigen-specific T cell responses and improve ongoing EAMG. The disease has also been prevented in Lewis rats by immunisation with a complementary peptide encoded by RNA complementary to the main immunogenic region of the AChR α-subunit (aa 67-76). Immunisation with such a complementary peptide, somewhat surprisingly, stimulated the production of anti-idiotypic antibodies that inhibited the recognition of native AChR by its antibodies, and hence prevented EAMG [65].

Conclusions Regarding Antigen-Specific Therapies in Animal Models

The results so far obtained in the field of antigen-specific immunomodulation of EAMG are encouraging. However, the dependence of the animal model on a restricted repertoire of T cell receptors and proteins involved in antigen presentation makes EAMG particularly suitable, and hence successful, for manipulation by AChR fragments, peptides and modified derivatives of the native antigen. By contrast, antigen-specific T helper cells from patients with MG recognise a large number of epitopes on the AChR molecule, but controversy still exists on the antigen used to detect such epitopes [66, 67], since antigen processing of synthetic peptides or recombinant fragments may be different from that of native AChR. On the basis of these considerations, antigen-specific immunotherapy of MG has been considered hardly feasible in the past. However, recent studies suggest that the repertoire of AChR epitopes is probably more restricted than previously thought [68, 69]. The identification of areas of preferential immunoreactivity on the AChR molecule suggests the need for further studies toward the possible application of such immunotherapeutic strategies to the human disease.

References

1. Marx A, Wilisch A, Schultz A et al (1997) Pathogenesis of myasthenia gravis. Virchows Arch 430:355-364
2. Lindstrom JM (2000) Acetylcholine receptors and myasthenia. Muscle Nerve 23:453-477
3. Engel AG (1994) Aquired autoimmune myasthenia gravis. In: Engel AG, Franzini-Armstrong G (eds) Myology, vol 2. McGraw-Hill, New York, pp 1769-1797
4. Grob D, Arsura E, Brunner N, Namba T (1987) The course of myasthenia gravis and therapies affecting outcome. Ann NY Acad Sci 505:472-499
5. Drachman D (1996) Immunotherapy in neuromuscular disorders: current and future strategies. Muscle Nerve 19:1239-1251
6. Blalock A (1944) Thymectomy in the treatment of myasthenia gravis. Report of 20 cases. J Thorac Surg 13:316-339
7. Jaretzki A III (1997) Thymectomy for myasthenia gravis. Analysis of the controversies regarding technique and results. Neurology 48 (Suppl 5):S52-S63
8. Scelsi R, Ferro MT, Scelsi L et al (1996) Detection and morphology of thymic remnants after video-assisted thoracoscopic extended thymectomy (VATET) in patients with myasthenia gravis. Int Surg 81:14-17
9. Mantegazza R, Confalonieri P, Antozzi C et al (1998) Video-assisted thoracoscopic extended thymectomy (VATET) in myasthenia gravis. Ann NY Acad Sci 841:749-752
10. Gronseth GS, Barohn RJ (2000) Practice parameter: thymectomy for autoimmune myasthenia gravis (an evidence-based review). Report of the Quality Standards Subcommittee of the American Academy of Neurology. Neurology 55:7-15
11. Sghirlanzoni A, Peluchetti D, Mantegazza R et al (1984) Myasthenia gravis: prolonged treatment with steroids. Neurology 34:170-174
12. Arsura E, Brunner NG, Namba T, Grob D (1985) High-dose intravenous methylprednisolone in myasthenia gravis. Arch Neurol 42:1149-1153

13. Cornelio F, Antozzi C, Mantegazza R et al (1993) Immunosuppressive treatments. Their efficacy on myasthenia gravis patients' outcome and on the natural course of the disease. Ann NY Acad Sci 681:594-602

14. Mertens HG, Hertel H, Reuther P, Ricker K (1981) Effect of immunosuppressive drugs (Azathioprine). Ann NY Acad Sci 377:691-699

15. Mantegazza R, Antozzi C, Peluchetti D et al (1988) Azathioprine as a single drug or in combination with steroids in the treatment of myasthenia gravis. J Neurol 235:449-453

16. Hohlfeld R, Michels M, Heininger K et al (1988) Azathioprine toxicity during long-term immunosuppression of generalized myasthenia gravis. Neurology 38:258-261

17. Gajdos P, Elkharrat D, Chevret A et al (1993) A randomized clinical trial comparing prednisone and azathioprine in myasthenia gravis. Results of the second interim analysis. J Neurol Neurosurg Psychiatry 56:1157-1163

18. Palace J, Newsom-Davis J, Lecky B (1998) A randomized double-blind trial of prednisolone alone or with azathioprine in myasthenia gravis. Myasthenia Gravis Study Group. Neurology 50:1778-1783

19. Hohlfeld R, Toyka KV, Besinger UA et al (1985) Myasthenia gravis: reactivation of clinical disease and of autoimmune factors after discontinuation of long-term azathioprine. Ann Neurol 17:238-242

20. Nyberg-Hansen N, Gjerstad L (1988) Myasthenia gravis treated with cyclosporin. Acta Neurol Scand 77:307-313

21. Goulon M, Elkharrat D, Gajdos P (1989) Treatment of severe myasthenia gravis with cyclosporin. A 12-month open trial. Presse Med 18:341-346

22. Tindall RS, Rollins JA, Phillips JT et al (1987) Preliminary results of a double-blind, randomized, placebo-controlled trial of cyclosporine in myasthenia gravis. N Engl J Med 316:719-724

23. Perez MC, Buot WL, Mercado-Danguilan C et al (1981) Stable remissions in myasthenia gravis. Neurology 31:32-37

24. Niakan E, Harati Y, Rolak LA (1986) Immunosuppressive drug therapy in myasthenia gravis. Arch Neurol 43: 155-156

25. Pinching AJ, Peters DK, Newsom-Davis J (1976) Remission of myasthenia gravis following plasma exchange. Lancet ii:1373-1376

26. Antozzi C, Gemma M, Regi B et al (1991) A short plasma exchange protocol is effective in severe myasthenia gravis. J Neurol 238: 103-107

27. Antozzi C, Berta E, Confalonieri P et al (1994) Protein-A adsorption is effective in immunosuppression resistant myasthenia gravis. Lancet 343:124

28. Berta E, Confalonieri P, Simoncini O et al (1994) Removal of antiacetylcholine receptor antibodies by protein A immunoadsorption in myasthenia gravis. Int J Artif Organs 17:455-460

29. Howard JF (1998) Intravenous immunoglobulins for the treatment of acquired myasthenia gravis. Neurology 51(Suppl 5):S30-S36

30. Gajdos P, Chevret S, Clair B et al (1997) Clinical trial of plasma exchange and high-dose intravenous immunoglobulin in myasthenia gravis. Myasthenia Gravis Clinical Study Group. Ann Neurol 41:789-796

31. Qureshi AI, Choudhry MA, Akbar S et al (1999) Plasma exchange and intravenous immunoglobulin treatment in myasthenic crisis. Neurology 52:629-632

32. Stricker RB, Kwiatkowska BJ, Habis JA, Kiprov DD (1993) Myasthenia gravis: response to plasmapheresis following failure of intravenous immunoglobulins. Arch Neurol 50:837-840

33. Christadoss P, Poussin M, Deng C (2000) Animal models of myasthenia gravis. Clin Immunol 94:75-87.

34. Lindstrom JM (2000) Acetylcholine receptors and myasthenia. Muscle Nerve 23:453-477

35. Tzartos S, Barkas T, Cung, T et al (1998) Anatomy of the antigenic structure of a large membrane antigen, the muscle type nicotinic acetylcholine receptor. Immunol Rev 163:89-120

36. Drachman DB, McIntosh KR, Yang B (1998) Factors that determine the severity of experimental myasthenia gravis. Ann NY Acad Sci 841:262-282

37. Christadoss P, Lindstrom J, Melvold R, Talal N (1985) I-A subregion mutation prevents experimental autoimmune myasthenia gravis. Immunogenetics 21:33-38

38. Kaul R, Shenoy M, Goluzko E, Christadoss P (1994) Major histocompatibility complex class II gene disruption prevents experimental autoimmune myasthenia gravis. J Immunol 152:3152-3157

39. Shenoy M, Goluzsko E, Christadoss P (1994) The pathogenic role of acetylcholine receptor α chain epitope within α146-162 in the development of experimental autoimmune myasthenia gravis in C57BL6 mice. Clin Immunol Immunopathol 73:338-343

40. Bellone M, Ostlie N, Lei S, Conti-Tronconi BM (1991) Experimental myasthenia gravis in congenic mice: sequence mapping and H-2 restriction of T helper epitopes on the α-subunits of Torpedo californica and murine acetylcholine receptor. Eur J Immunol 21:2303-2310

41. Fujii Y, Lindstrom J (1998) Specificity of the T cell immune response to acetylcholine receptor in experimental autoimmune myasthenia gravis: response to subunits and synthetic peptides. J Immunol 140:1830-1837

42. Drachman DB (1996) Immunotherapy in neuromuscular disorders: current and future strategies. Muscle Nerve 19:1239-1251

43. Killen J, Lindstrom J (1984) Specific killing of lymphocytes which cause EAMG by ricin acetylcholine receptor conjugates. J Immunol 133:2549-2553

44. McIntosh KR, Linsley PS, Bacha PA, Drachman DB (1998) Immunotherapy of experimental myasthenia gravis: selective effects of CTLA4Ig and synergistic combination of IL-2-diphtheria toxin fusion protein. J Neuroimmunol 87:136-146

45. Wu J-M, Wu B, Guarnieri F et al (2000) Targeting antigen specific T cells by genetically engineered antigen presenting cells. A strategy for specific immunotherapy of autoimmune disease. J Neuroimmunol 106:145-153

46. Weiner HL (1997). Oral tolerance: immune mechanisms and treatment of autoimmune diseases. Immunol Today 18:335-343

47. Wang ZY, Qiao J, Link H (1993) Suppression of experimental autoimmune myasthenia gravis by oral administration of acetylcholine receptor. J Neuroimmunol 44: 209-214

48. Okumura S, McIntosh K, Drachman DB (1994) Oral administration of acetylcholine receptor: effects on experimental autoimmune myasthenia gravis. Ann Neurol 36:704-713

49. Ma CG, Zhang GX, Xiao BG et al (1995) Suppression of experimental autoimmune myasthenia gravis by nasal administration of acetylcholine receptor. J Neuroimmunol 58:51-60

50. Drachman DB, Okumura S, Adams RN, McIntosh K (1996) Oral tolerance in myasthenia gravis. Ann N Y Acad Sci 778:258-272

51. Barchan D, Souroujon M, Im S-H et al (1999) Antigen specific modulation of experimental myasthenia gravis: nasal tolerization with recombinant fragments of the human acetylcholine receptor a subunit. Proc Natl Acad Sci USA 96:8086-8091

52. Im S-H, Barchan D, Fuchs S, Souroujon MC (1999) Suppression of ongoing experimental myasthenia by oral treatment with an acetylcholine receptor recombinant fragment. J Clin Invest 104:1723-1730

53. Im S-H, Barchan D, Souroujon MC, Fuchs S (2000) Role of tolerogen conformation in induction of oral tolerance in experimental autoimmune myasthenia gravis. J Immunol 165:3599-3605

54. Bartfeld D, Fuchs S (1973) Specific immunosuppression of experimental myasthenia gravis by denatured acetylcholine receptor. Proc Natl Acad Sci USA 75:4006-4010

55. Wu B, Deng C, Goluszko E, Christadoss P (1997) Tolerance to a dominant T cell epitope in the acetylcholine receptor molecule induces epitope spread and suppresses murine myasthenia gravis. J Immunol 159:3016-3023

56. Karachunski PI, Ostlie NS, Okita DK et al (1999) Subcutaneous administration of T-epitope sequences of the acetylcholine receptor prevents experimental myasthenia gravis. J Neuroimmunol 93:108-121

57. Karachunski PI, Ostlie NS, Okita DK, Conti-Fine BM (1997) Prevention of experimental myasthenia gravis by nasal administration of synthetic acetylcholine receptor T epitope sequences. J Clin Invest 100:3027-3035

58. Baggi F, Andreetta F, Caspani E et al (1999) Oral administration of an immunodominant T-cell epitope downregulates Th1/Yh2 cytokines and preventes experimental myasthenia. J Clin Invest 104:1287-1295

59. Katz-Levy Y, Kirshner SL, Sela M, Mozes E (1993) Inhibition of T-cell reactivity to myasthenogenic epitopes of the human acetylcholine receptor by synthetic analogs. Proc Natl Acad Sci USA 90:7000-7004

60. Katz-Levy Y, Paas-Rozner M, Kirshner S et al (1997) A peptide composed of tandem analogs of two myasthenogenic T cell epitopes interferes with specific autoimmune responses. Proc Natl Acad Sci USA 94:3200-3205

61. Katz-Levy Y, Dayan M, Wirguin I et al (1998) Single amino acid analogs of a myasthenogenic peptide modulate specific T cell responses and prevent the induction of experimental myasthenia gravis. J Neuroimmunol 85:78-86

62. Faber-Elmann A, Paas-Rozner M, Sela M, Mozes E (1998) Altered peptide ligands act as partial agonists by inhibiting phospholipase C activity induced by myasthenogenic T cell epitopes. Proc Natl Acad Sci USA 95:14320-14325

63. Paas-Rozner M, Dayan M, Paas Y et al (2000) Oral administration of a dual analog of two myasthenogenic T cell epitopes down-regulates experimental autoimmune myasthenia gravis in mice. Proc Natl Acad Sci USA 97:2168-2173

64. Spack EG, McCutcheon M, Corbelletta N et al (1995) Induction of tolerance in experimental autoimmune myasthenia gravis with solubilized MHC classII: acetylcholine receptor peptide complexes. J Autoimmun 8:787-807

65. Araga S, LeBoeuf RD, Blalock JE (1993) Prevention of experimental autoimmune myasthenia gravis by manipulation of the immune network with a complementary peptide for the acetylcholine receptor. Proc Natl Acad Sci USA 90:8747-8751

66. Protti MP, Manfredi AA, Horton RM et al (1993) Myasthenia gravis: recognition of a human autoantigen at the molecular level. Immunol Today 14:363-368

67. Hawke S, Matsuo H, Nicolle M et al (1996) Autoimmune T cells in myasthenia gravis: heterogeneity and potential for specific immunotargeting. Immunol Today 17:307-311

68. Wang ZY, Okita DK, Howard J, Conti-Fine BM (1997) Th1 epitope repertoire on the alpha subunit of human muscle acetylcholine receptor in myasthenia gravis. Neurology 48:1643-1653

69. Hill M, Beeson D, Moss P et al (1999) Early-onset myasthenia gravis: a recurring T-cell epitope in the adult-specific acetylcholine receptor epsilon subunit presented by the susceptibility allele HLA-DR52a. Ann Neurol 45:224-231

Chapter 5

Ontogeny of Skeletal Muscle Cells

G. Cossu

Introduction

The differentiation of skeletal muscle is an early and crucial step in the development of vertebrates since it provides the embryo with motility in the early stages. Skeletal myogenesis begins shortly after gastrulation but persists, at least in mammals, until the end of postnatal growth, and the potential for myogenesis continues for the entire life span of the animal [1]. Local signalling commits mesodermal cells to a myogenic fate, and shortly afterwards they begin to synthesise contractile proteins that accumulate in the cytoplasm and self-assemble into sarcomeres. Motility is dependent upon shortening of the sarcomeres, paracrystalline structures that are specialised for transforming chemical energy into movement. The advantage of accumulating millions of sarcomeres within a single cytoplasm has led to multinucleation, a different strategy from the coupling of single cells adopted by the heart. Within the highly structured cytoplasm of the multinucleated muscle fibre mitosis is no longer possible, and when experimentally induced by oncogenes it leads to disruption of the spindle and death (mitotic catastrophe). As a consequence, growth of the muscle fibre during fetal and postnatal development depends upon the addition of single cells, which must be instructed on when to divide and when to differentiate, by fusing either with pre-existing fibres or among themselves to generate a new fibre. It is therefore obvious that diversification of myogenic cell fate is as crucial as their commitment. It allows production of postmitotic skeletal muscle during early embryogenesis, and at the same time maintains a pool of mitotic progenitors that permit further growth of the tissue as required and the ability to regenerate in response to injury.

In this review I will discuss current knowledge of the early steps of skeletal myogenesis, and possible mechanisms that ensure maintenance of a progenitor pool during later development.

Stem Cell Research Institute, San Raffaele Scientific Institute, DIBIT, Via Olgettina 58, 20132 Milan; Department of Histology and Medical Embryology, "La Sapienza" University, Via A. Scarpa 14, 00185 Rome, Italy. e-mail: giulio.cossu@hsr.it; giulio.cossu@uniroma1.it

Myogenic Commitment: Signals from Neighbouring Tissues

The skeletal muscles of the vertebrate body are known to derive from the dorsal domain of the somites, spheres of paraxial mesoderm that form in a craniocaudal succession along the neural tube/notochord axis [2]. Cells located in the ventral domain of the somites, the sclerotome, on the other hand, will form cartilage and bone. At the onset of somitogenesis, cells are not yet determined; dorsal/ventral or medial/lateral rotation of the epithelial somite does not perturb subsequent development, suggesting that it is signals from the environment that determine the identity of cells within the newly formed somite.

Axial structures (neural tube/notochord complex) are required to promote myogenesis, but only in the precursors of epaxial (back) muscles located in the dorsomedial domain of newly formed somites. By contrast, the precursors of hypaxial (limb and body wall) muscles, located in the dorsolateral half of the somites, require a signal from the dorsal ectoderm for myogenic commitment (reviewed in [3]). Cells from the medial domain activate *Myf5* when cultured in the presence of axial structures, while cells from the lateral half activate *MyoD* when cultured with their own dorsal ectoderm [4]. Thus, in mammals, axial structures activate epaxial myogenesis through a *Myf5*-dependent pathway, while dorsal ectoderm acts on hypaxial progenitors through a *MyoD*-dependent pathway. The latter is also dependent upon previous expression of either *Myf5* or *Pax3* [5]. Subsequently, the great majority of myogenic cells express both *MyoD* and *Myf5*, although with variable intensity. This explains why *Myf5*-null embryos have primarily epaxial muscle defects, whereas *MyoD*-null embryos have predominantly hypaxial muscle defects [6, 7]. In both cases, the residual gene is sufficient to support almost normal skeletal muscle development throughout the body.

In the lateral myogenic progenitor cells, *MyoD* expression and subsequent terminal differentiation is transiently repressed in order to allow migration to the limb and body wall, where muscle formation will take place. Although irreversibly committed (as shown by classic transplantation experiments), these myogenic progenitors do not express any member of the *MyoD* family either in the somite or during migration [8]. They can be identified by the expression of the transcription factor *Pax3* [9] and the receptor tyrosine kinase *c-met* [10]. It is therefore likely that their differentiation is repressed by signals derived from adjacent tissues, and BMP4 has been shown to be able to mimic the inhibitory effect of lateral mesoderm [11].

Possible Role of *Shh* and *Wnt* in the Activation of Myogenesis

The notochord produces a ventralising signal that activates *Pax1* and specifies a sclerotomal fate: *Sonic hedgehog* (*Shh*), normally produced by the notochord, can mimic this activity [12]. However, *Shh* is also required to promote myogenesis, and indeed in *Shh*-null embryos, epaxial myogenesis is absent, whereas progenitors of hypaxial myogenesis are specified normally [13]. On the other hand, the neural tube produces several members of the *Wnt* family that can activate the myogenic program in the dorsal part of the somite [14].

As discussed above, the onset of *MyoD* and *Myf5* expression is spatially and temporally regulated in mouse embryos. The action of the neural tube in activating *Myf5* can be replaced by cells expressing *Wnt1*, while *MyoD* activation by dorsal ectoderm can be replaced by *Wnt7a*-expressing cells [15]. *Wnt7a* is expressed in the correct spatiotemporal pattern to be a candidate molecule for this activity. *Shh* synergises with both *Wnt1* and *Wnt7a*, even though it is not expressed in the ectoderm.

Together these data allow a simple model of signalling activity to be proposed, as shown in Fig. 1. *Shh*, produced by the notochord and floor plate, activates *Pax1*

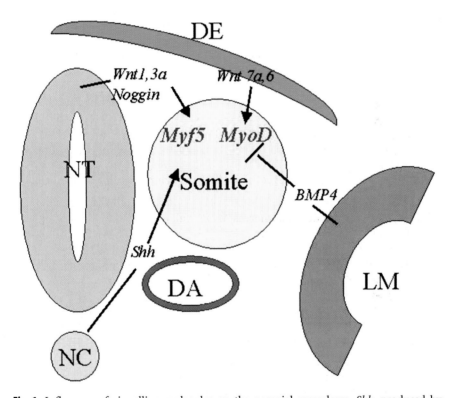

Fig. 1. Influences of signalling molecules on the paraxial mesoderm. *Shh*, produced by notochord (NC) and floor plate, acts on the ventral domain of newly formed epithelial somites, inducing sclerotome, and also on the dorsomedial domain, inducing medial dermomyotome. *Wnt1* (and *Wnt3a*), produced by the dorsal neural tube (NT) acts (with *Shh*) on the dorsomedial domain of newly formed somites, where *Myf5* expression is soon after observed and epaxial progenitors are specified. *Wnt7a* (and possibly *Wnt6*), produced by dorsal ectoderm (DE), acts on the dorsolateral domain, where hypaxial progenitors are specified. *BMP4*, produced by lateral mesoderm (LM), prevents *MyoD* activation and early differentiation in the lateral domain of somites. Its action is counteracted by direct binding of *Noggin*, produced by the dorsal neural tube. DA, dorsal aorta; DE, dorsal ectoderm; LM lateral mesoderm; NT, neural tube; NC, notochord; *Shh*, Sonic hedgehog; *BMP*, bone morphogenetic protein

and, in conjunction with *BMP4* [16], leads to chondrogenesis in the future sclerotome. *Shh*, in conjunction with *Wnt1* (and possibly other *Wnts*), activates myogenesis in the future dermomyotome, via a *Myf5*-dependent pathway. Different *Wnts* such as *Wnt7a* activate myogenesis in the lateral domain, probably through a *MyoD*-dependent pathway. This activity is inhibited by *BMP4* to prevent premature differentiation; the negative action of *BMP4* is counteracted, probably through direct protein-protein interactions with *noggin*, which is produced by the dorsal neural tube in a Wnt-dependent manner [17, 18]. While further data on new molecules will probably add to the complexity of this current model, more stringent evidence, such as in situ inhibition by specific antibodies (many of which are not yet available) and detailed analysis of mutant embryos, will be required to confirm its validity.

From Receptors to Downstream Genes

While *BMP* and *noggin* are likely to interact directly, *Wnt* and *Shh* act through classic membrane receptors. In the case of *Wnts*, vertebrate homologues of *Drosophila Frizzled* are considered as putative receptors [19], and so far about 10 members of this family have been cloned in different organisms [20, 21]. Many of these are expressed in somites (*Fz1, 3, 7, 8, 9*) with distinct but partially overlapping patterns. For example, *Fz1* is expressed along the medial border, consistent with a possible preferential interaction with *Wnt1* from the adjacent dorsal neural tube. On the other hand, *Fz7* is expressed in a pattern complementary to *Fz1*, i.e., along the lateral and caudal edge of newly formed somites, consistent with the possibility of a preferential interaction with *Wnt7a*. *Wnt1* acts through the classic Dishevelled → GSK3 → β-catenin pathway; by contrast, *Wnt7a* appears to act through a β-catenin-independent pathway [22] and leads to *MyoD* rather than to *Myf5* activation [15]. It is thus tempting to speculate that *Fz1* and *Fz7* may mediate the differential response to *Wnt1* and *Wnt7a* and activate different intracellular pathways.

Another possible level of regulation for *Wnt* signaling may be exerted by the sFRPs (*Related Proteins Frizzled* soluble), a new class of genes recently identified in several laboratories [23, 24]. These are secreted molecules with a strong homology with the *Frizzled* extracellular domain. Among those examined, only *Frzb1* was found to be expressed in the presomitic mesoderm and newly formed somites. *Frzb1* totally inhibits somitic myogenesis but has no effect on more mature somites or on myogenic cell lines, and thus appears to act differently from intracellular myogenic inhibitors such as Id or Twist. Genes downstream of the *Wnt* signalling pathway such as *En1*, *Noggin*, and *Myf5* were down-regulated, but *Pax3* and *Mox1* were not, excluding a generalised toxic effect [25]. These results are in keeping with results from the *Wnt1-Wnt3a* double knockout and corroborate the idea that *Wnt* signals may act by regulating both myogenic commitment and expansion of committed cells in the mouse mesoderm, while *Frzb1* would counteract this activity in most premyogenic areas. Indeed, in mouse embryos lacking both *Wnt1* and *Wnt3a*, the medial compartment of the dermomyotome is not formed and the expression of *Myf5* is decreased [26] but not abolished, probably because partial activation by *Shh* has already occurred.

From the data discussed above, it appears that, at least medially, *Shh* and *Wnts* cooperate to activate myogenesis, but may also instruct diversification between epaxial and hypaxial myogenesis. How this can be achieved in molecular terms is still far from clear. *Shh* binds to *Patched*, a receptor that activates an intracellular pathway involving protein kinase A and ultimately leading to activation of several zinc finger proteins, termed Gli. Activated Gli may bind directly to regulatory regions of *My5* and *MyoD* promoters, the complexity of which has made these studies difficult. Similarly, Tcf/β-catenin complex, activated by *Wnt1* (see below), may directly activate transcription of target genes but may also contribute to open the *Myf5* locus, making it more easily accessible to other transcription factors. It should however be remembered that *Shh* has been reported to be an important survival factor for paraxial mesoderm [27] and to have mitogenic activity on myoblasts [28]. Several *Wnts*, including *Wnt1* and *Wnt7a*, have strong mitogenic and often transforming activity and most likely act as survival factors as well [20]. Thus, both molecules have the potential to activate, directly or indirectly, transcription of *Myf5* and *MyoD*, and also to promote survival and expansion of the committed population. In reality, a combinatorial action of *Shh/Wnt* transcriptional activation, proliferation and survival must account for the final number of differentiated cells in a given structure such as the myotome. This is relevant to the next issue discussed in this review, namely, how different fates are chosen within contiguous and probably equivalent cells of the epithelial dermomyotome [29].

Generation of Myoblast Diversity and the Origin of Different Fibers

Most of the work discussed above on the activation of myogenesis relates to the formation of the myotome, the first patterned array of terminally differentiated, postmitotic, mononucleated muscle cells. It is still unknown whether commitment of the progenitors of later phases of myogenesis occurs through the same mechanisms and in the same spatiotemporal context. As a matter of fact, activation of the myogenic program leads to terminal differentiation of a fraction of somitic cells; the remaining cells are probably kept in a committed but undifferentiated state. These are the embryonic and fetal myoblasts that will produce primary and secondary fibres respectively [30], and perhaps also satellite cells during later development. In *Drosophila*, lateral inhibition through *Notch* and *Delta* has been shown to be the probable mechanism by which adult myogenic progenitors are selected in response to *Wng* signalling [31]. It thus appears likely that a similar mechanism may operate in the mammalian somite. Indeed, several *Delta* and *Notch* isoforms are expressed in the somites [32], and *Notch* inhibits myogenesis, probably through different intracellular mechanisms [33, 34]. However, direct evidence for a role of *Notch* in diversifying cell fate in mammalian somites is still missing.

Receptors for growth factors may be pertinent targets for *Notch* signalling. It has been proposed that the dorsal portion of the neural tube inhibits terminal myogenic differentiation through production of growth factors [35]; therefore, some kind of heterogeneity may be invoked to explain the differential fate of myotomal precursors versus other precursors, similar to that observed between embryonic and fetal myoblasts in the developing limb bud (see below). From this point of view it is interesting to note that the neural tube produces various fibroblast growth factors (FGFs) [36], and the first cells which form the myotome are the only myogenic cells which do not express the FGF receptor *FREK* [37]. Whether a preselected myogenic population fails to respond to FGF, or whether local signalling prevents expression of FGF receptors in a homogeneous population, is unclear since the location of myotomal precursors in the dermomyotome is still unknown.

On a teleological basis, this can be explained by the need to maintain a precursor pool of dividing myogenic cells (to cope with the growing size of the embryo) and at the same time to generate differentiated, postmitotic skeletal muscle fibres (to allow early movements of the embryo). A possible mechanism to ensure that certain myoblasts will differentiate in an environment that is permissive for proliferation may be based on the inability of these myoblasts to respond to growth factors and/or to molecules which inhibit differentiation. A few years ago, we proposed a possible mechanism by which TGFβ might influence the process of primary fibre formation in vivo. Committed myoblasts will proliferate in the presence of mitogens and will differentiate in their absence [1]. It is therefore conceivable that a gradient of mitogen concentration is established throughout the proximal-distal axis of the growing limb, with the lowest concentration present at the base of the limb, where primary fibres initially form. Myoblasts will proliferate in the growing distal extremity of the limb bud, where the concentration of mitogens is high (progress zone), and will differentiate first at the base of the limb bud, distant from the source of mitogens. However, the limb also contains a high concentration of TGFβ that inhibits myogenesis. Inasmuch as embryonic myoblasts are insensitive to TGFβ [38], they may escape this inhibition and differentiate into primary fibres. Fetal myoblasts, on the other hand, may remain undifferentiated until the onset of a new wave of proliferation promoted by the growth factors released by primary fibers and necessary to generate the pool of precursor cells for the formation of secondary fibres.

This model has the advantage of reconciling previously unexplained data such as the presence of growth factors in newly formed muscle with the differential effect of TGFβ on embryonic and fetal myoblasts [38]. It was later proposed that selective expression of the θ isoform of protein kinase C in fetal myoblasts would mediate the inhibitory effect of TGFβ on these cells [39]. However, this hypothesis, shown in Fig. 2, still awaits in vivo functional analysis with dominant positive and negative TGFβ signalling molecules.

Satellite Cells

Satellite cells are classically defined as quiescent mononucleated cells located between the sarcolemma and the basal lamina of adult skeletal muscle [40]. They

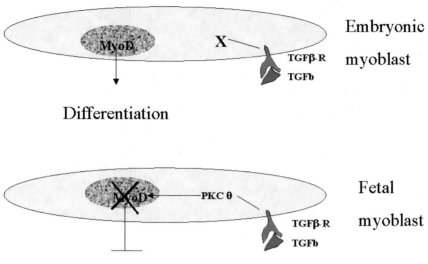

Fig. 2. A simplified model describing fate diversification in embryonic and fetal myoblasts needed to generate primary and secondary fibres. Embryonic myoblasts, which do not express PKCθ are insensitive to TGFβ and differentiate in the absence of mitogens. Fetal myoblasts, which express PKCθ, are inhibited by TGFβ and remain available as a pool of undifferentiated cells that will later resume proliferation in response to growth factors produced by newly formed primary fibres

contribute to postnatal growth of muscle fibres, whose nuclei cannot divide. At the end of longitudinal growth, satellite cells become quiescent but can be activated if the existing fibres are damaged or destroyed. In this case they undergo a number of cells divisions, producing fusion-competent cells that can either fuse with damaged fibres or form new ones, and other cells that return to quiescence, thus maintaining a progenitor pool. This fact has led to the suggestion that they represent a type of stem cells [41, 42].

Previous work identified specific features of satellite cells (morphology, resistance to phorbol esters but susceptibility to TGFβ-induced block of differentiation, early expression of acetylcholine receptors and acetylcholinesterase) that characterise them as a different class of myogenic cells with respect to embryonic and fetal myoblasts (reviewed in [43]). Satellite cells are first detectable when the basal lamina forms in vivo. In this period, intensive myoblast fusion occurs, leading to a drastic reduction of myogenic mononucleated cells. Since satellite cells do not undergo differentiation at this time, the control of proliferation and differentiation in these cells must be different, so as to allow the persistence of mononucleated undifferentiated myogenic cells in the postnatal and in the adult muscle.

Are Myogenic Cells Only Derived From Satellite Cells in Regenerating Muscle?

Satellite cells are the only relatively well-defined myogenic cell in postnatal life. It is currently assumed but not experimentally proven that they represent a single cell type with a common embryological origin. The origin of satellite cells is presumed to be somitic, but the evidence for this is not conclusive due to technical difficulties in identifying quail nuclei in chick-quail chimeras at the ultrastructural level [44]. Furthermore, it is surprising that despite a low number of resident quiescent satellite cells (identified by their location and the expression of *M-Cadherin*) in adult healthy muscle, hundreds of activated (*MyoD*-positive) satellite cells are seen hours after an injury to the tissue [45]. This suggests that cells are recruited to muscle regeneration from additional sources, either locally or systemically.

A number of observations have pointed to the unorthodox appearance of muscle cells in a variety of tissues or cell culture systems that were neither myogenic nor derived from somites. For example, spontaneous myogenic differentiation of cells from the brain has been repeatedly documented, but only through insertion of the reporter gene *LacZ* into the *Myf5* locus was it possible unequivocally to identify *Myf5*-expressing cells in the nervous system and to show that these cells coexpress neural and muscle markers [46]. Hence, even though there is no clue as to the physiological significance of these findings, they provide an indication for potential myogenic precursors in sites other than muscle.

Similarly, several laboratories have shown that primary fibroblasts from different organs are able to undergo muscle differentiation at significant frequency when co-cultured with myogenic cells (reviewed in [47]). It is important to emphasise that during early embryogenesis, myogenesis occurs in the absence of pre-existing muscle and in proximity to signalling centers. By contrast, fetal and postnatal myogenesis occurs far away from signaling centres (many of which have disappeared) and in close proximity to developing muscle fibres. It is thus not unreasonable to speculate that the latter now represent the source of signals that recruit cells from the mesoderm to myogenesis (Fig. 3).

A search for donor tissues that may contribute myogenic cells for muscle regeneration identified bone marrow as a possible source. By transplanting genetically marked bone marrow into immune-deficient mice, it was shown that marrow-derived cells may undergo myogenic differentiation and, moreover, that they can circulate [48]. These results raised questions concerning the origin of these circulating myogenic progenitors and their possible relationship with resident satellite cells. The large majority of clones with the typical morphology of mouse adult satellite cells were derived from dorsal aorta and not from somites, the presumed source of all skeletal myogenic cells of the body. In vitro, these aorta-derived myogenic cells express a number of myogenic and endothelial markers that are also expressed by satellite cells. In vivo, aorta-derived myogenic progenitors participate in muscle regeneration and fuse with resident satellite cells [49]. These data suggest that a subset of postnatal satellite cells may be rooted in a vascular lineage. Whether these myogenic vascular cells arise from a primordial per-

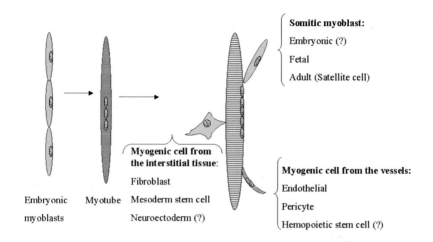

Fusion into myotubes   Growth of the muscle fiber by addition of:

(no muscle present)   (muscle already present)

Fig. 3. Possible origin of myogenic cells in the fetal and perinatal periods of development. While in the embryo muscle fibres forms by fusion of embryonic myoblasts in the absence of pre-existing muscle and in response to signals from surrounding tissues (see Fig. 1), later in development myogenic cells are added to developing muscle. In this case, these cells are far away from the original signalling centres (many of which no longer exist) and close to developing muscle fibres. It is reasonable to assume that they are recruited by signals released from muscle itself. In this case their origin does not need to be somitic, since commitment occurs locally. Possible other origins (interstitial tissue and vessels) are indicated

icyte or from endothelial cells proper, as suggested by the expression of endothelial markers, is not currently known. Furthermore, these cells may be pluripotent since clones of dorsal aorta can give rise to osteoblast-like cells in the presence of *BMP2*; indeed pluripotency is preserved even in adult muscle satellite cells, as proven by the fact that *BMP2* can switch them to an osteogenetic fate [50].

When ingressing a developing muscle anlage, these progenitors should find themselves in a muscle field and thus adopt a satellite cell fate (mimicked by appropriate culture conditions); when the vasculature develops inside a different tissue, these cells may adopt the specific fate of that tissue and contribute to its histogenesis (Fig. 4). The only tissue in which these progenitors may remain easily accessible (because of its loose stroma) may be the bone marrow, and this would explain our previous observation [48]. Pluripotent mesenchymal cells capable of producing osteoblasts, chondroblasts, adipocytes and even skeletal muscle have long been known to be present in the bone marrow and are a subject of intense investigation; however, bone marrow stromal cells do not express endothelial markers [51]. Whether the cells originating from embryonic vessels represent the progenitors of pluripotent mesenchymal cells, or a separate lin-

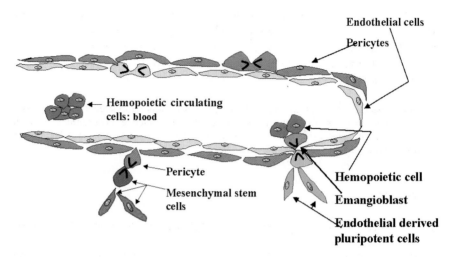

Fig. 4. Model explaining the origin of pluripotent cells. A growing vessel, comprised of an endothelial layer, surrounding pericytes and circulating hemopoietic cells penetrates into a developing tissue. Asymmetric mitosis, both in the endothelial and in the perithelial layer, may generate cells that leave the vessel and adopt the fate of the tissue (e.g., muscle) that the vessel has entered. They may differentiate into myoblasts or remain as undifferentiated progenitors (i.e. satellite cells)

eage with at least part of the same developmental potential, remains to be investigated.

Conclusions

As in many other fields of vertebrate developmental biology, information is accumulated through series of experiments that utilise similar approaches. It is almost too obvious to say that analysis of mutant mice will help to unravel the process of commitment, diversification, and pluripotentiality of mammalian skeletal myogenic cells. Indeed, recent history has shown that only a careful combination of different approaches may be really informative. The complexity of a single mutant phenotype is often such that years of work are required to understand it. For example, not until 10 years after the first report do we have a possible explanation of why the *Myf5*-null mice lack ribs, where the gene is never expressed [52]. In this context, simpler approaches, such as organ cultures that allow experimental manipulation of the tissues, often combined with now available genetic markers, have contributed significantly to the unravelling of complex biological phenomena such as, for example, myogenic commitment. The information available now should permit us to go back to different mutant embryos and ask whether or not a certain inductive event will occur in the absence of a given gene. In the case of the origin of cells at later stages, both classic chimeric studies and lineage analysis with retroviral or genetic labelling (e.g., the *cre-lox* system) will allow the examination in vivo of the fate of progenitors originating from a given structure. Even with the caveat of these systems, it is conceivable that within the next 5 years we may be able to understand in much greater detail the early steps of mammalian myogenesis.

Acknowledgements. The work in the authors' laboratory was supported by grants from Telethon, the European Community, Human Frontiers Science Programme (HFSP), Fondazione Istituto-Pasteur Cenci Bolognetti, Ministero Università e Ricerca Scientifica (MURST) and Agenzia Spaziale Italiana (ASI).

References

1. Hauschka SD (1994) The embryonic origin of skeletal muscle. In: The scientific basis of myology. Academic Press, pp 3-72
2. Christ B, Ordhal CP (1994) Early stages of chick somite development. Anat Embryol 191:381-396
3. Cossu G, Tajbakhsh S, Buckingham M (1996) Myogenic specification in mammals. Trends Genet 12:218-223
4. Cossu G, Kelly R, Tajbakhsh S et al (1996) Activation of different myogenic pathways: Myf5 is induced by the neural tube and MyoD by the dorsal ectoderm in mouse paraxial mesoderm. Development 122:429-437
5. Tajbakhsh S, Rocancourt D, Cossu G, Buckingham M (1997) Redefining the genetic hierarchies controlling skeletal myogenesis: Pax-3 and Myf5 act upstream of MyoD. Cell 89:127-138
6. Braun T, Rudnicki MA, Arnold HH, Jaenisch R (1992) Targeted inactivation of the muscle regulatory gene Myf5 results in abnormal rib development and perinatal death. Cell 71:369-382
7. Kablar B, Krastel K, Ying C et al (1997) MyoD and Myf5 differentially regulate the development of limb versus trunk skeletal muscle. Development 124:4729-4738
8. Tajbakhsh S, Buckingham ME (1994) Mouse limb muscle is determined in the absence of the earliest myogenic factor Myf5. Proc Natl Acad Sci USA 91:747-751
9. Bober E, Franz T, Arnold HH et al (1994) Pax-3 is required for the development of limb muscles: a possible role for the migration of dermomyotomal muscle progenitor cells. Development 120:603-612
10. Bladt F, Riethmacher D, Isenmann S et al (1995) Essential role for the c-met receptor in the migration of myogenic precursor cells into the limb bud. Nature 376:768-771
11. Pourquié O, Fan CM, Coltey M et al (1996) Lateral and axial signals involved in avian somite patterning: a role for BMP4. Cell 84:461-471
12. Fan C, Tessier-Lavigne M (1994) Patterning of mammalian somites by surface ectoderm and notochord: evidence for sclerotome induction by a hedgehog homolog. Cell 79:1175-1186
13. Borycki AG, Brunk B, Tajbakhsh S et al (1999) Sonic hedgehog control epaxial muscle deteminao Myf5 activation. Development 126:4053-4063
14. Münsterberg AE, Kitajewski J, Bumcroft DA et al (1995) Combinatorial signaling by Sonic hedgehog and Wnt family members induces myogenic bHLH gene expression in the somite. Genes Dev 9:2911-2922
15. Tajbakhsh S, Borello U, Vivarelli E et al (1998) Differential activation of Myf5 and

MyoD by different Wnts in explants of mouse paraxial mesoderm and the later activation of myogenesis in the absence of Myf5. Development 125:4155-4162

16. Murtaugh LC, Chyung JH, Lassar AB (1999) Sonic hedgehog promotes somitic chondrogenesis by altering the cellular response to BMP signaling. Genes Dev 15:225-237

17. Hirsinger E, Duprez D, Jouve C et al (1997) Noggin acts downstream of Wnt and Sonic Hedgehog to antagonize BMP4 in avian somite patterning. Development 124:4605-4614

18. Marcelle C, Stark MR, Bronner-Fraser M (1997) Coordinate actions of BMPs, Wnts, Shh and noggin mediate patterning of the dorsal somite. Development 124:3955-3963

19. Banhot P, Brink M, Samos CH et al (1996) A new member of the frizzled family from Drosophila functions as a wingless receptor. Nature 382:225-230

20. Wodarz A, Nusse R (1998) Mechanisms of Wnt signaling in development. Annu Rev Cell Dev Biol 14:59-88

21. Dierick H, Bejsovec A (1999) Cellular mechanisms of wingless/Wnt signal transduction. Curr Top Dev Biol 43:153-190

22. Kengaku M, Capdevila J, Rodriguez-Esteban C et al (1998) WNT3a regulates AER formation and utilizes an intracellular signaling pathway distinct from the dorso-ventral signal WNT7a during chick limb morphogenesis. Science 280:1274-1277

23. Leyns L, Bouwmeester T, Kim S-H et al (1997) Frzb-1 is a secreted antagonist of Wnt signaling expressed in the Spemann organizer. Cell 88:747-756

24. Wang S, Krinks M, Lin K et al (1997) Frzb, a secreted protein expressed in the Spemann organizer, binds and inhibits Wnt-8. Cell 88:757-766

25. Borello U, Coletta M, Tajbakhsh S et al (1999) Trans-placental delivery of the Wnt antagonist Frzb1 inhibits development of caudal paraxial mesoderm and skeletal myogenesis in mouse embryos. Development 126:4247-4255

26. Ikeya M, Takada S (1998) Wnt signaling from the dorsal neural tube is required for the formation of the medial dermomyotome. Development 125:4969-4976

27. Teillet M-A, Watanabe Y, Jeffs P et al (1998) Sonic hedgehog is required for survival of both myogenic and chondrogenic somitic lineages. Development 125:2019-2030

28. Duprez D, Fournier-Thibault C, Douarin N le (1998) Sonic hedgehog induces proliferation of committed skeletal muscle cells in the chick limb. Development 125:495-505

29. Tajbakhsh S, Cossu G (1997) Establishing myogenic identity during somitogenesis. Curr Opin Genet Dev 7:634-641

30. Kelly AM, Zachs S (1969) The histogenesis of rat intercostal muscle. J Cell Biol 42:154-169

31. Baylies MK, Bate M, Ruiz Gomez M (1998) Myogenesis: a view from Drosophila. Cell 93:921-927

32. McGrew MJ, Pourquié O (1998) Somitogenesis: segmenting a vertebrate. Curr Opin Genet Dev 8:487-493

33. Wilson-Rawls J, Molkentin JD, Black BL, Olson EN (1999) Activated Notch inhibits myogenic activity of the MADS-Box transcription factor myocyte enhancer factor 2C. Mol Cell Biol 4:2853-2862

34. Nofziger D, Miyamoto A, Lyons KM, Weinmaster G (1999) Notch signaling imposes two distinct blocks in the differentiation of C2C12 myoblasts. Development 126:1689-1702

35. Buffinger N, Stockdale PB (1994) Myogenic specification in somites: induction by axial structures. Development 120:1443-1452

36. Kalcheim C, Neufeld G (1990) Expression of basic fibroblast growth factor in the nervous system of early avian embryos. Development 109:203-215

37. Marcelle C, Wolf J, Bonner-Fraser M (1995) The in vivo expression of the FGF recep-

tor FREK mRNA in avian myoblasts suggests a role in muscle growth and differentiation. Dev Biol 172:100-114

38. Cusella de Angelis MG, Molinari S, Ledonne A et al (1994) Differential response of embryonic and fetal myoblasts to TGFβ: a possible regulatory mechanism of skeletal muscle histogenesis. Development 120:925-933

39. Zappelli F, Willems D, Osada S et al (1996) The inhibition of differentiation caused by TGFβ in fetal myoblasts is dependant upon selective expression of PKCθ: A possible molecular basis for myoblast diversification during limb histogenesis. Dev Biol 180:156-164

40. Bischoff R (1994) The satellite cell and muscle regeneration, In: Engel AG, Franzini-Armstrong C (eds) Myology, 2nd edn. McGraw-Hill, New York, pp 97-133

41. Miller JB, Schaefer L, Dominov JA (1999) Seeking muscle stem cells. Curr Top Dev Biol 43:191-219

42. Seale P, Rudnicki MA (2000) A new look at the origin, function, and "stem-cell" status of muscle satellite cells. Dev Biol 218:115-124

43. Cossu G, Molinaro M (1987) Cell heterogeneity in the myogenic lineage. Curr Top Dev Biol 23:185-208

44. Armand O, Boutineau AM, Mauger A et al (1983) Origin of satellite cells in avian skeletal muscles. Arch Anat Microsc 72:163-181

45. Grounds MD, Garrett KL, Lai MC et al (1992) Identification of skeletal muscle precursor cells in vivo by use of MyoD1 and myogenin probes. Cell Tissue Res 267:99-104

46. Tajbakhsh S, Vivarelli G, Cusella de Angelis G et al (1994) A population of myogenic cells derived from the mouse neural tube. Neuron 13:813-821

47. Cossu G (1997) Unorthodox myogenesis: possible developmental significance and implications for tissue histogenesis and regeneration. Histol Histopathol 12:755-760

48. Ferrari G, Cusella de Angelis MG, Coletta M et al (1998) Skeletal muscle regeneration by bone marrow derived myogenic progenitors. Science 279:1528-1530

49. De Angelis L, Berghella L, Coletta M et al (1999) Skeletal myogenic progenitors originating from embryonic dorsal aorta co-express endothelial and myogenic markers and contribute to post-natal muscle growth and regeneration. J Cell Biol 147:869-878

50. Katagiri T, Yamaguchi A, Komaki M et al (1994) Bone morphogenetic protein-2 converts the differentiation pathway of C2C12 myoblasts into the osteoblast lineage. J Cell Biol 127:1755-1766

51. Prockop DJ (1997) Marrow stromal cells as stem cells for nonhematopoietic tissues. Science 276:71-74

52. Kaul A, Koster M, Neahus H, Braun T (2000) Myf-5 revisited: loss of early myotome formation does not lead to a rib phenotype in homozygous Myf-5 mutant mice. Cell 102:17-19

Chapter 6

Idiopathic Inflammatory Myopathies: Immunological Aspects

R. Mantegazza[1], P. Bernasconi[1], F. Cornelio[2]

Muscle inflammation is generally termed "myositis" whether the aetiology is known (viral, bacterial or parasitic) or unknown (idiopathic). The inflammatory myopathies are a heterogeneous group of subacute/chronic muscle disorders sharing a common characteristic muscle degeneration mediated by inflammatory mechanisms [1]. This review will be concerned with the main pathogenetic features of the three major inflammatory myopathies: dermatomyositis (DM), polymyositis (PM) and inclusion body myositis (IBM). The latter includes sporadic (s-IBM) and hereditary inclusion body myopathy (h-IBM), which is an a hereditary progressive muscle disease with muscle pathology similar to the s-IBM, but lacking lymphocytic inflammation [2].

PM and DM are considered to be autoimmune diseases on the basis of the following features: muscle damage at the endomysial level by infiltrating T cells in PM and complement-mediated humoral attack against endothelial cells in DM; frequent association with other autoimmune diseases; serum positivity for autoantibodies and positive responses to immunosuppressive treatment [3-7]. Figure 1 provides a simplified representation of the main pathogenetic mechanisms of PM/IBM and DM. With respect to s-IBM, it is still unclear whether the immune response is a primary or a secondary event [8, 9].

Inflammatory Features

The immunological characteristics of muscle cells in the inflammatory myopathies are summarised in Table 1. The main characteristic of polymyositis and s-IBM is a mononuclear cell infiltrate, mainly composed of CD8+ T cells and macrophages, which surrounds and invades single non-necrotic muscle fibres located in the endomysium [10]. The CD8+ T cells are in the activated state since they are HLA-DR+ [10] and LFA-1+ [11] and have a memory phenotype (CD45RO+) [12]. The antigens that initiate and trigger the immune reaction are

[1] Myopathology and Immunology Unit, Department of Neuromuscular Diseases; [2] Department of Neuromuscular Diseases, National Neurological Institute "Carlo Besta", Via Celoria 11, 20133 Milan, Italy. e-mail: rmantegazza@istituto-besta.it

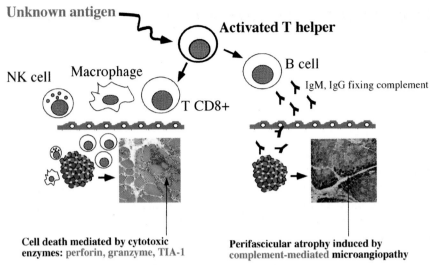

Fig. 1. The possible mechanisms leading to muscle derangement in inflammatory myopathies are illustrated. Left-hand side of picture refers to PM/IBM degeneration as showed in hematoxilin/eosin stained muscle biopsy. Muscle damage is mainly due to the action of CD8+ T lymphocytes via release of cytotoxic mediators. Right-hand side of picture referres to DM degeneration as showed in NADH-stained muscle biopsy. Perifascicular atrophy may result from the action of immunoglobulins (IgM and IgG) fixing complement secreted by B lymphocytes. The trigger, still unknown, induce activation and recruitment of specific T helper lymphocytes into muscle tissue. *NK cell*, natural killer cell

still unknown. In s-IBM, non-necrotic muscle fibres are more frequently invaded than fibres containing the characteristic congophilic inclusions (see below), suggesting the importance of an immune-mediated mechanism in the aetiology of this disease [13].

In dermatomyositis, the complement-mediated deposition of the lytic membrane attack complex on endomysial capillaries induces perivascular inflammation, capillary depletion, muscle fibre necrosis and atrophy [14, 15]. Infiltrating cells are predominantly CD4+ T cells and B cells [3].

MHC Class I and Class II and Co-stimulatory Molecule Expression on Muscle

Major histocompatibility complex (MHC) class I and class II molecules are involved in antigen presentation to immune-competent T cells. MHC class I molecules are constitutively expressed on most cell types but are low or absent on normal muscle fibres, while MHC class II are expressed only on professional antigen-presenting cells (APC) or other cell types after adequate stimulation [16]. In the inflammatory myopathies, MHC class I and II may be variably expressed (Tables 1, 2).

Table 1. Immunobiological features of effector cells in inflammatory myopathies

	Dermatomyositis	Polymyositis	s-IBM
T lymphocytes			
Endomysial	+	+++	++
Perimysial	++	++	+
Perivascular	+/++	+	+
B lymphocytes			
Endomysial	+	+/-	+/-
Perimysial	++	+	+
Perivascular	+++	+	+/-
Macrophages	+	+/++	+/++
NK cells	+/-	+	+
T cell receptor			
α/β heterodimer	Present	Present	Present
γ/δ heterodimer	Rare	Rare	Rare
α/β repertoire[a]	Polyclonal	Oligoclonal	Oligoclonal
CDR3	Random	Conserved	Random/conserved
Cytokines			
IL-1α	PA	In EC and IC	In EC and IC
IL-1β	In MC	In MC	In MC
TNF-α	+/-	+/-	++
IL-2/IL-2R	+	+	-
IFN-γ	+/-	+	+
IL-4/5/6/10/13	-	+/-	+/-
TGF-β1	++	+	Not done
Chemokines			
MCP-1	++	+	+
MIP-1α	+	++	++
MIP-1β	+	+	+
Matrix metalloproteinases			
MMP-2	-	-	-
MMP-9	+	+	+
Cytotoxic enzymes			
Perforin	Absent	Present	Present
Granzyme	Absent	Present	Present
Apoptotic signals			
Fas	-/+	+	+
FasL	-/+	++	++
Stimulatory signals			
CD40L	CD4+>CD8+	CD4+>CD8+	Not done
CD28/CTLA4	-	T cells invading BB-1+ muscle fibers	T cells invading BB-1+ muscle fibers

[a] Polyclonal or oligoclonal TCR repertoire expressed as number of Vα or Vβ rearrangements detected.
IBM, Inclusion body myositis; *EC*, endothelial cells; *IC*, infiltrating cells; *PA*, perifascicular arterioles; *MC*, mononuclear cells
+++, strong signal; ++, medium signal; +, weak signal; -/+, very weak signal; -, no signal

Table 2. Immunobiological features of target cells in inflammatory myopathies

	Dermatomyositis	Polymyositis	s-IBM
MHC			
Class I	+++	+++	++
Class II	++	++	+
Adhesion molecules			
ICAM-1	–	+++	+++
VCAM-1	–	–	–
LFA-1	–	–	–
LFA-3	–	+	+
Co–stimulatory molecules			
B7.1	–	–	–
B7.2	–	–	–
BB-1	++ (only on regenerating N-CAM+ fibres)	+++	+++
Stimulatory signals			
CD40	+	+	Not done
CD40L	–	–	Not done
Matrix metalloproteinases			
MMP-2	+ (only on regenerating fibres)	+	+
MMP-9	+ (only on regenerating fibres)	+	+
Apoptotic signals			
Fas	–/+	+	+
Bcl-2	not done	+	+

IBM, inclusion body myositis
+++, strong signal; ++, medium signal; +, weak signal; –/+, very weak signal; –, no signal

In polymyositis and s-IBM, early, widespread overexpression of MHC class I has been observed on muscle fibres [17]. This includes not only muscle cells invaded by cellular infiltrates, but also muscle fibres at a distance from the inflammation. This suggests that MHC class I expression may be independent of, and may precede, the release of pro-inflammatory cytokines by infiltrating cells. Very recently, MHC class I expression has been up-regulated, via a controllable muscle-specific promoter system, in skeletal muscles of young mice, which then developed an inflammatory myositis resembling polymyositis [18]. This up-regulation, in turn, activated the expression of other molecules such as intercellular adhesion molecule-1 (ICAM-1) (on muscle fibres and muscle-infiltrating cells), macrophage inflammatory protein (MIP) -1α, monocyte chemoattractant protein (MCP)-1, and interleukin-15 (detected in muscle biopsies by RNase protection assay). There was induction of serum levels of autoantibodies against histidyl-tRNA synthetase, the most common autoantigen detected in PM [18]. On the basis of these results, the authors speculated that chronic up-regulation of MHC class I, perhaps in patients with certain genetic backgrounds, is able to induce a self-sustained inflammatory response, and that the context, location and duration of the stimulus may be as important as its specificity (e.g. viral infection) [18].

In dermatomyositis, muscle endothelial cells are positive for MHC class II expression, in association with a predominance of B lymphocytes at perivascular sites [3].

Cultured myoblasts and myotubes constitutively express MHC class I and lymphocyte function-associated molecule 3 (LFA-3), but are able to increase expression or express de novo MHC class I, MHC class II (DR>DP>DQ) or ICAM-1 after stimulation with IFN-γ or IFN-γ plus TNF-α [19-25]. When cultured myoblasts fuse to multi-nucleated myotubes, MHC class I expression disappears, suggesting a role of MHC class I/II (possibly also ICAM-1) in myogenesis [22].

Naive T cells need to be stimulated by (a) the interaction between T cell receptor and antigen-MHC complex, and (b) the interaction between CD28, a molecule expressed on their surface, and the "co-stimulatory molecules", B7.1 or B7.2, expressed on the target cells [26]. To further support the hypothesis that muscle fibres play a role in T cell activation in inflammatory myopathy [27], expression of co-stimulatory molecules has been investigated on muscle cells in vivo and in vitro. B7.1 and B7.2 were never detected in vivo on inflammatory myopathy muscle fibres and in vitro on cultured myoblasts, even when treated by pro-inflammatory stimuli [28]. However, another co-stimulatory molecule distinct from B7.1 and B7.2, named BB-1 [29], has been found to be strongly expressed on many muscle fibres of polymyositis and s-IBM patients, with a distribution similar to that of MHC class I molecules, but not on those from non-myopathic muscles [30, 31]. These results suggest that muscle fibres might actively participate in the immune response via synthesis of BB-1 in the inflammatory myopathies; this molecule might also be active in the inflammatory reactions that occur after conventional intramuscular vaccination or myoblast transfer for gene therapy [32]. However, in a more extensive study, the authors re-evaluated the role of BB-1. They found that a minor fraction of in vitro cultured myoblasts stimulated with IFN-γ and/or TNF-α do express BB-1 molecule on their surface; this may indicate heterogeneity of the myoblasts or a genetic variability in BB-1 expression [33]. A summary of co-stimulatory molecule expression on muscle cells is given in Table 2.

T Cell Receptor Gene Rearrangement

T cells recognize the antigen presented by MHC class I and II molecules via their antigen-specific T cell receptor (TCR), an α/β (or γ/δ) heterodimer derived from rearrangements of the variable (V), diversity (D), joining (J) and constant (C) segments of the TCR gene [34]. The TCR contact region with an antigenic peptide is called the complementarity-determining region 3 (CDR3) and represents products of V-J and V-D-J segment rearrangements for the α and β TCR chains respectively [35]. Analysis of CDR3 sequences, and Vα and Vβ usage, can provide information indicative of a role for the involvement of the antigen or superantigen in T cell recruitment within the inflamed area.

We and other groups analysed the TCR repertoire of infiltrating T lymphocytes in muscle biopsies from patients with inflammatory myopathy by polymerase chain reaction (PCR) and/or immunocytochemistry. In polymyositis, but not in dermatomyositis, oligoclonality of $V\alpha$ and $V\beta$ TCR families, and restricted CDR3 usage with a consensus motif in the CDR3 region were found [36-38]. By double-fluorescence immunocytochemistry, Bender et al. demonstrated that 70% of $V\beta13.1^+$ and $V\beta3^+$ T lymphocytes were $CD8^+$ muscle-invading lymphocytes, with a conserved consensus sequence in the CDR3 region. These findings suggest that $CD8^+$ T cells are recruited into the inflamed areas of polymyositis muscles by the action of a conventional, but unknown, antigen, most likely expressed on the muscle cell surface. To date, only one polymyositis patient has been observed whose muscle-infiltrating T lymphocytes were of the γ/δ TCR subset with the unusual phenotype $V\gamma3\text{-}J\gamma1\text{-}C\gamma1/V\delta2\text{-}J\delta3\text{-}C\delta$ [39]. Although O'Hanlon et al. [40] found that 20% of 42 inflammatory myopathy patients were positive for the γ/δ TCR rearrangement, no similarities between clonotypes at the junctional regions of the γ/δ TCR were found, suggesting that γ/δ T lymphocytes are likely to be of secondary importance in the myocytotoxic process in the inflammatory myopathies.

Several groups have investigated the TCR repertoire of muscle-infiltrating T cells in s-IBM. An oligoclonal TCR repertoire with heterogeneity in the CDR3 domain has been found [41-45], suggesting that, in s-IBM, T cell recruitment into muscle tissue may be stimulated by a superantigen or by superantigen-like activity. However, Bender et al. [46], combining PCR and immunocytochemistry procedures, showed that $CD8^+$ $V\beta5.3^+$ T cells invaded muscle fibres while $CD8^-$ $V\beta5.1^+$ T cells were distributed perivascularly. Sequence analysis of CDR3 revealed that $CD8^+$ $V\beta5.3^+$ T cells were clonally restricted while $CD8^-$ $V\beta5.1^+$ cells were diverse [46]. Recently, Dalakas et al. analysed multiple muscle biopsies of s-IBM patients over a period of 19-22 months and found that restricted expression of $V\beta$ families and the CDR3 region among T cells persisted over the course of the disease, even at the later stages when there is a significant loss of fibres [8]. These results suggest that, even if immunosuppressive treatment in IBM is not effective, the immune system plays a crucial part in the pathogenesis of the disease. A summary of TCR characterisation as observed in the inflammatory myopathies is given in Table 1.

Cytokine and Chemokine Expression

Cytokines play a crucial role in inflammatory reactions. These molecules are soluble, short-lived proteins produced, constitutively or only with specific stimulation, by several cell types [47]. In inflammatory myopathies, several studies have looked at the degree and distribution of cytokine expression [48-54], although a clear description of the cytokine expression in muscle tissue in inflammatory myopathy has not yet been established (see Tables 1 and 2). In general, predominant expression of interleukin (IL)-1α, IL-1β and TNF-α has been found by immunological and molecular studies. In particular, IL-1α was detected in the endothelial cells of endomysial capillaries and in perifascicular arterioles and venules, and in the infiltrating mononuclear cells, whereas IL-1β was limited to mononuclear cells. TNF-α was occasionally expressed in mononuclear cells and on muscle fibre membranes. Both Th1-type and Th2-type cytokines were investigated: IL-2 and IFN-γ were found in some inflammatory myopathy patients, but usually only in a few muscle cells; IL-4, -5, -6, -10 and -13 were detected only at very low levels. The discrepancies among several studies might be due to the different procedures used or to the fact that these molecules are synthesised early in the inflammatory event before the patients undergo muscle biopsy.

In inflammatory myopathy muscle biopsies, in particular in dermatomyositis, TGF-β1 was immunolocalised to the extracellular matrix and never in close proximity to mononuclear cell infiltrates [55]. TGF-β1 is a pleiotropic cytokine: it acts as a pro- and anti-inflammatory molecule, promotes fibrosis on damaged tissues, and exhibits regenerating and anti-angiogenic functions [56]. In dermatomyositis cases, the overexpression of TGF-β1 might be due to the ischaemic muscle damage characteristic of this disorder [55]. Moreover, TGF-β1, linked to the extracellular matrix, might contribute to the recruitment of mononuclear cells within the muscle, since it increases the adhesiveness of endothelial cells for leukocytes, inhibits E-selectin expression in endothelial cells [57] and induces MCP-1 synthesis [58].

MCP-1, like MIP-1α, MIP-1β and RANTES, belongs to the family of β-chemokines, chemotactic proteins that induce the directional migration of several cell types, including neutrophils, monocytes and T lymphocytes, into immunologically active sites [59]. The β-chemokines (also known as "C-C chemokines", due to the position of two cysteine residues adjacent to each other), together with the α-chemokines, are the largest and best-characterised families, and function primarily as activators and chemoattractors of T and B lymphocytes and monocytes [59]. MCP-1 attracts monocytes, memory T cells and natural killer cells [60, 61]. MIP-1α is a potent chemoattractor of CD8+ T cells [62]; while MIP-1β and RANTES preferentially attract CD4+ T cells. Thus, these chemokines might play an active role in the pathogenesis of inflammatory myopathies. Up to now, few studies have investigated the β-chemokine expression in muscle tissue from inflammatory myopathy by PCR and immunohistochemistry: MCP-1, MIP-1α, MIP-1β and RANTES transcripts were detected in almost all muscle biopsies from inflammatory myopathy but not in control muscles; β-chemokines were found in all inflammatory myopathic muscle sections localised to infiltrating inflammatory cells and the extracellular matrix, with a pattern of distribution related to the different pathogenetic processes underlying the three forms of inflammatory myopathy [63-65]. In polymyositis and IBM muscle biopsies, MIP-1α was more widely distributed than MCP-1, while in dermatomyositis muscle, MCP-1 predominated over MIP-1α [64].

Muscle cells might be actively involved in β-chemokine synthesis and release in the inflamed area. Recently, we demonstrated that IL-8, a C-X-C chemokine, and RANTES were constitutively expressed by cultured myoblasts and enhanced after pro-inflammatory stimuli; both MCP-1 transcription and its release in culture supernatants were induced by IFN-γ or TNF-α stimulus [66]. These data further support the hypothesis that muscle cells are not only the target of the immune-mediated attack, but that they may directly release cytokines and/or chemokines necessary to initiate and perpetuate immunocompetent cell recruitment.

CD40-CD40 Ligand Interaction

CD40 and CD40 ligand (CD40L) are molecules involved in B cell proliferation, differentiation and stimulation of IgG production, as well as in cytokine production and up-regulation of several cell surface molecules on non-lymphoid cells. Increased CD40 expression was found on muscle fibres from polymyositis and dermatomyositis, and conversely CD40L was expressed on muscle-infiltrating T cells, mainly CD4+ but also CD8+ [67]. In vitro, CD40 ligation stimulates the production of inflammatory cytokines (e.g. IL-6, IL-15) and chemokines (in particular MCP-1 and IL-8) by human myoblasts. CD40 expression on myoblasts is induced in a dose-dependent manner by IFN-γ and TNF-α treatment, two pro-inflammatory cytokines reported to be present in polymyositis/dermatomyositis muscle tissues [48-51, 54, 67]. These results indicate that infiltrating T cells induce muscle cells to secrete cytokines and chemokines via CD40-CD40L interaction, thus perpetuating the chronic inflammatory state observed in muscle tissues in the inflammatory myopathies. CD40-CD40L expression on effector and target cells is summarised in Tables 1 and 2.

Matrix Metalloproteinases

Matrix metalloproteinases (MMPs) are calcium-dependent zinc endoproteinases involved in the remodelling of the extracellular matrix in physiological and pathological conditions [68]. In the latter, MMPs have different activities. They degrade matrix components, induce macrophage and T cell adhesion to matrices and endothelial cells and their recruitment from blood stream into tissues, propagate an inflammatory response and deposite amyloid proteins [69]. In the inflammatory myopathies, MMP expression has been investigated by immunocytochemistry, and the activity of MMPs has been tested by gelatin substrate zymography [70] (see Tables 1 and 2). In all inflammatory myopathy muscle biopsies studied, MMP-2 and MMP-9 were found in the endomysial blood vessels and in regenerating muscle fibres. In polymyositis and s-IBM only, MMP-2 and MMP-9 were also found on the membranes of muscle fibres invaded by CD8+ T cells in association with MHC class I molecules [70]. This finding supports an active role for MMPs in promoting T cell adhesion to muscle, and T cell cytotoxicity, by degrading the basal lamina which surrounds muscle fibres. Moreover, MMP-2 co-localised with the β-amyloid precursor protein (β-APP) within the vacuoles in s-IBM, suggesting that a chronic inflammatory state induces overexpression of MMPs, which in turn enhances the expression of β-amyloid, thus creating a vicious circle.

T Cell- and Complement-Mediated Cytotoxicity

The degeneration and necrosis of muscle fibres by CD8+ T cells in PM and s-IBM is predominantly mediated by release of cytotoxic enzymes, such as perforin and granzyme [71]. Transcripts for these molecules were detected in polymyositis muscle biopsies [48, 72]. Perforin+ T cells were found in association with invaded non-necrotic muscle fibres [72-74]. Perforin was concentrated in the contact area between invading T cells and the muscle fibres, suggesting that these T cells are specifically activated by the muscle. When dermatomyositis muscle biopsies were analyzed, perforin was found randomly distributed in the cytoplasm of T cells, indicating a non-specific T cell activation [72]. A summary of the distribution of cytotoxic enzymes is given in Table 1.

Much effort has been devoted to understanding the role of the Fas-FasL interaction in inducing cell death in inflammatory myopathy [75-77]. In polymyositis and s-IBM many muscle fibres express Fas on their surface, and muscle-infiltrating T cells express FasL [75, 76]; however, no sign of apoptosis have been observed in the MHC class I expressing muscle fibres [77] (see Table 2). Fas-FasL interactions do not seem to play a crucial role in myocytotoxicity in the inflammatory myopathies.

In dermatomyositis, deposition of C5b-9 complement membrane attack complex (MAC) on small blood vessels precedes the infiltration of inflammatory cells, mainly B lymphocytes, CD4+ T cells and macrophages, within the perivascular site. Moreover, MAC deposition might be responsible for structural abnormalities of muscle such as perifascicular atrophy and muscle fibre necrosis [14, 15, 78]. The trigger of this complement activation is still unknown.

Vacuoles in s-IBM

A striking feature of s-IBM is the abnormal accumulation of a variety of different proteins within muscle fibres: β-APP, prion protein, ubiquitin, α1-antichymotrypsin, phosphorylated tau, apolipoprotein E and low-density lipoprotein receptors, TGF-β1, neuronal and inducible nitric oxide synthases, nitrotyrosine and superoxide dismutase-1 [79]. All these proteins, except for phosphorylated tau, are accumulated in normal human muscles only at the postsynaptic domain of the neuromuscular junction. Askanas et al. proposed the presence of a "junctionalizing master gene" which modulates transcription of the proteins of the neuromuscular junction: in s-IBM this master gene could be activated by a virus, while, in h-IBM it could be mutated [79]. Recently, hyperexpression of p42 mito-gen-activated protein kinase (MAPK) and of a 35-kDa phosphoprotein, a protein likely to originate from the cytoskeleton, were found in vacuolated muscle fibres of s-IBM patients and not in other inflammatory or vacuolar myopathies [80]. These data suggest an ongoing cytoskeletal disruption with accumulation of phosphorylated microtubular proteins. The events that lead to this phenomenon – i.e. pro-inflammatory cytokines, cytotoxic enzymes, MMPs – are unclear.

Clinical Features

PM patients are generally older than 20 years of age; PM is rare in childhood [81]. Incidence and prevalence are about 0.6/100 000 and 6.3/100 000, respectively, and the female:male ratio is about 3:1. Proximal muscles are more involved than dis-tal muscles, and the patients suffer from myalgia and muscle tenderness. Oropharyngeal and esophageal muscles are also involved, causing dysphagia in one third of PM cases [82]. No cutaneous manifestations or involvement of facial or ocular muscles are present in PM. Serum creatine kinase (CK) levels are ele-vated (5- to 50-fold). Antinuclear antibodies (ANA) are detected in 16%-40% of PM patients; antibodies to Jo-1 and to signal recognition proteins are common in PM and probably identify a subgroup of patients [82].

s-IBM is the most common muscle disease in patients older than 50 years of age and shows a male predominance [9]. In IBM weakness develops more insid-iously than in the other forms of IM, and the evolution of the disease is very slow, thus delaying diagnosis. In IBM, the distal muscles are more involved than the proximal muscles, with characteristic weakness of the long finger flexors and knee extensors. Muscle involvement is often asymmetric [9]. Mild facial weak-ness can be observed but extraocular muscles are not compromised. The slow progression of the disease usually leads to severe disability and eventually to res-piratory muscle weakness. Dysphagia occurs in up to 60% of cases, especially late in the disease [9, 82] and swallowing difficulties are common [83]. CK and aldolase levels are normal in about 20%-30% of cases and moderately increased in the majority. ANA, anti Jo-1 and antibodies to signal recognition proteins are rarely detected in s-IBM [84, 85]. At light microscopy, muscle biopsy shows mononuclear cell infiltrates, groups of atrophic fibres, muscle fibres with eosinophilic cytoplasmic inclusions and rimmed vacuoles containing granular material, an increase in ragged-red fibres and in cytochrome c oxidase (COX)-negative muscle fibres. Staining with Congo red, thioflavine S, or crystal violet indicates that vacuoles within muscle fibres contain amyloid [2]. At electron microscopy, vacuolated muscle fibres show 15- to 21-nm diameter intranuclear or cytoplasmic paired helical filaments and 6- to 10-nm amyloid-like fibrils accu-mulated in the cytoplasm [2]. In patients affected by the hereditary form of IBM most vacuolated muscle fibres are not positive for Congo red staining, while ragged-red fibres and COX-negative muscle fibres are not present. Within vacuo-lated muscle fibres, paired helical filaments do not have some epitopes of phosphorylated tau, are not congophilic, and do not contain apolipoprotein E and cellular prion protein [2].

DM can occur at any age – in infancy [3], childhood (5-14 years) [81], or adulthood. The incidence and prevalence are similar to those of PM, and so is the female:male ratio. Symptoms mainly involve skin and muscle tissue. Gottron's rash is constituted by slightly raised violaceous papules and plaques and covers mainly metacarpophalangeal joints, proximal interphalangeal joints, and distal interphalangeal joints [86]. The heliotrope rash is distributed symmetrically on the periorbital skin. Malar erythema, poikiloderma on exposed skin, violaceous erythema on the extensor surfaces, periungual and cuticular changes can also be present [86]. Skin lesions may precede the development of myopathy and may persist after control of myopathy is achieved. Proximal muscles are usually affected and distal muscles may be involved later [3]. Initial symptoms are usually myalgia, fatigue, or weakness; in advanced cases facial muscles can be involved, but extraocular muscles are usually spared. Bulbar muscle involvement induces dysphagia in approximately 30% of patients and indicates rapid progression of the disease, associated with a poor diagnosis [3, 86]. In up to 40% of children or adolescents with DM, but unusually in adults, calcinosis of skin or muscle can be observed [3]. Muscle damage correlates with an increase of serum levels of CK, aldolase, myoglobin, lactate dehydrogenase, aspartate aminotransferase, and alanine aminotransferase [82]. In DM patients' sera, ANA can be detected, and antibodies to Mi-2, a 240-kDa nuclear protein, can be found in 15%-25% of DM patients. However, although this test is specific for the disease it is not sensitive [82, 86]. Antibodies to Jo-1 and Ro (SS-A) are rarely detected in DM [86]. The diagnosis of DM relies mainly on muscle biopsy, which shows the characteristic perifascicular atrophy of muscle fibres in up to 90% of children and in 50% of adults [3].

Treatments

Immunosuppressive treatments are the main therapeutic approaches in DM and PM. A high proportion of DM and PM patients respond positively to treatment with corticosteroids alone or in combination with another immunosuppressive drug. In the majority of cases a complete or worthwhile functional remission is achieved [87]. However, in some cases there is a resistance to therapy. In these cases the initial treatment regimen seems to be important in determining the clinical course and outcome, i.e., intravenous methylprednisolone rather than oral prednisone [87], and combinations of immunosuppressive drugs, e.g., azathioprine and prednisone [87, 88]. High-dose intravenous immunoglobulin (IVIg) therapy has been investigated in open and controlled trials in IM patients [88]. These results justify the use of IVIg in patients who respond poorly or fail to respond to corticosteroids and immunosuppressive drugs. IVIg should be considered for juvenile DM, due to the harmful effects of corticosteroids in growing children [88]. IBM is still difficult to treat; immunosuppressive or immunomodulating drugs have limited effects [89]. Open or double-blind, placebo-controlled trials on the effect of IVIg in IBM did not show significant benefits [88]. IVIg combined with prednisone for a 3-month period was not effective [90].

Conclusions

Understanding of the immunopathogenesis of inflammatory myopathies has been greatly improved by application of techniques in molecular and cellular immunology. Although the many observations that have been made have not yet led to a better understanding of the aetiologies of these disorders, they have opened new possibilities for therapeutic interventions which may be better tolerated by patients suffering from these severe diseases.

References

1. Dalakas MC (1998) Molecular immunology and genetics of inflammatory muscle diseases. Arch Neurol 55:1509-1512
2. Askanas V, Serratrice G, Engel WK (eds) (1998) Inclusion-body myositis and myopathies. Cambridge University Press, Cambridge
3. Engel AG, Hohlfeld R, Banker BQ (1994) Inflammatory myopathies. The polymyositis and dermatomyositis syndromes. In: Engel AG, Franzini-Armstrong C (eds) Myology. McGraw-Hill, New York, pp 1335-1383
4. Ioannou Y, Sultan S, Isenberg D (1999) Myositis overlap syndromes. Curr Opin Rheumatol 11:468-474
5. Miller FW (1993) Myositis-specific autoantibodies. Touchstones for understanding the inflammatory myopathies. JAMA 270:1846-1849.
6. Mantegazza R, Baggi F, Bernasconi P (2001) Methods for examination of antibodies in skeletal muscle disease. In: Preedy VR, Peters TJ (eds) Skeletal muscle: pathology, diagnosis and management of disease. Greenwich Medical Media, London (in press)
7. Basta M, Dalakas MC (1994) High dose intravenous immunoglobulin exerts its beneficial effect in patients with dermatomyositis by blocking endomysial deposition of activated complement fragments. J Clin Invest 94:1729-1735
8. Amemiya K, Granger RP, Dalakas MC (2000) Clonal restriction of T-cell receptor expression by infiltrating lymphocytes in inclusion body myositis persists over time: studies in repeated muscle biopsies. Brain 123:2030-2039
9. Cherin P (1999) Treatment of inclusion body myositis. Curr Opin Rheumatol 11:456-461
10. Engel AG, Arahata K (1984) Monoclonal antibody analysis of mononuclear cells in myopathies. II: phenotypes of autoinvasive cells in polymyositis and inclusion body myositis. Ann Neurol 16:209-215
11. De Bleecker JL, Engel AG (1994) Expression of cell adhesion molecules in inflammatory myopathies and Duchenne dystrophy. J Neuropathol Exp Neurol 53:369-376
12. De Bleecker JL, Engel AG (1995) Immunocytochemical study of CD45 T cell isoforms in inflammatory myopathies. Am J Pathol 146:1178-1187
13. Pruitt JN, Showalter CJ, Engel AG (1996) Sporadic inclusion body myositis: counts of different types of abnormal fibers. Ann Neurol 39:139-143

14. Kissel JT, Mendell JR, Rammohan KW (1986) Microvascular deposition of complement membrane attack complex in dermatomyositis. N Engl J Med 314:331-334
15. Emslie-Smith AM, Engel AG (1990) Microvascular changes in early and advanced dermatomyositis: a quantitative study. Ann Neurol 27:343-356
16. Germain RN, Margulies DH (1993) The biochemistry and cell biology of antigen processing and presentation. Annu Rev Immunol 11:403-450
17. Emslie-Smith AM, Arahata K, Engel AG (1989) Major histocompatibility complex class I antigen expression, immunolocalization of interferon subtypes, and T cell-mediated cytotoxicity in myopathies. Hum Pathol 20:224-231
18. Nagaraju K, Raben N, Loeffler L et al (2000) Conditional up-regulation of MHC class I in skeletal muscle leads to self-sustaining autoimmune myositis and myositis-specific autoantibodies. Proc Natl Acad Sci USA 97:9209-9214
19. Mantegazza R, Hughes SM, Mitchell D et al (1991) Modulation of MHC class II antigen expression in human myoblasts after treatment with IFN-γ. Neurology 41:1128-1132
20. Holhfeld R, Engel AG (1990) Induction of HLA-DR expression on human myoblasts with interferon-gamma. Am J Pathol 136:503-508
21. Goebels N, Michaelis D, Wekerle H, Hohlfeld R (1992) Human myoblasts as antigen-presenting cells. J Immunol 149:661-667
22. Mantegazza R, Gebbia M, Mora M, Barresi R, Bernasconi P, Baggi F, Cornelio F (1996) Major histocompatibility complex class II molecule expression on muscle cells is regulated by differentiation: implications for the immunopathogenesis of muscle autoimmune diseases. J Neuroimmunol 68:53-60
23. Beauchamp JR, Abraham DJ, Bou-Gharios G, Partridge TA, Olsen I (1992) Expression and function of heterotypic adhesion molecules during differentiation of human skeletal muscle in culture. Am J Pathol 140:387-401
24. Michaelis D, Goebels N, Hohlfeld R (1993) Constitutive and cytokine-induced expression of human leukocyte antigens and cell adhesion molecules by human myotubes. Am J Pathol 143:1142-1149
25. Hardiman O, Faustman D, Li X, Sklar RM, Brown RH (1993) Expression of major histocompatibility complex antigens in cultures of clonally derived human myoblasts. Neurology 43:604-608
26. Reiser H, Stadecker MJ (1996) Costimulatory B7 molecules in the pathogenesis of infectious and autoimmune diseases. N Engl J Med 335:1369-1377
27. Hohlfeld R, Engel AG (1994) The immunobiology of muscle. Immunol Today 15:269-274
28. Bernasconi P, Confalonieri P, Andreetta F, Baggi F, Cornelio F, Mantegazza R (1998) The expression of co-stimulatory and accessory molecules on cultured human muscle cells is not dependent on stimulus by pro-inflammatory cytokines: relevance for the pathogenesis of inflammatory myopathy. J Neuroimmunol 85:52-58
29. Freeman GJ, Cardoso AA, Boussiotis VA et al (1998) The BB1 monoclonal antibody recognizes both cell surface CD74 (MHC class II-associated invariant chain) as well as B7-1 (CD80), resolving the question regarding a third CD28/CTLA-4 counterreceptor. J Immunol 161:2708-2715
30. Beherens L, Kerschensteiner M, Misgeld T, Goebels N, Wekerle H, Hohlfeld R (1998) Human muscle cells express a functional costimulatory molecule distinct from B7.1 (CD80) and B7.2 (CD86) in vitro and in inflammatory lesions. J Immunol 161:5943-5951
31. Murata K, Dalakas MC (1999) Expression of the costimulatory molecule BB-1, the ligands CTLA-4 and CD28, and their mRNA in inflammatory myopathies. Am J Pathol 155:453-460

32. Blau HM, Springer ML (1995) Muscle-mediated gene therapy. N Engl J Med 333:1554-1556

33. Beherens L, Kerschensteiner M, Misgeld T, Goebels N, Wekerle H, Hohlfeld R (1998) Human muscle cells express a functional costimulatory molecule distinct from B7.1 (CD80) and B7.2 (CD86) in vitro and in inflammatory lesions. J Immunol 2000 164:5330 correction of ref 30

34. Garcia KC, Teyton L, Wilson IA (1999) Structural basis of T cell recognition. Ann Rev Immunol 17:369-397

35. Davis MM, Boniface JJ, Reich Z et al (1998) Ligand recognition by $\alpha\beta$ T cell receptors. Ann Rev Immunol 16:523-544

36. Mantegazza R, Andreetta F, Bernasconi P et al (1993) Analysis of T cell receptor repertoire of muscle-infiltrating T lymphocytes in polymyositis. J Clin Invest 91:2880-2886

37. Bender A, Ernst N, Iglesias A, Dornmair K, Wekerle H, Hohlfeld R (1995) T cell receptor repertoire in polymyositis: clonal expansion of autoaggressive CD8+ T cells. J Exp Med 181:1863-1868

38. O'Hanlon TP, Dalakas MC, Plotz PH, Miller FW (1994) Predominant TCR-$\alpha\beta$ variable and joining gene expression by muscle-infiltrating lymphocytes in the idiopathic inflammatory myopathies. J Immunol 152: 2569-2576

39. Pluschke G, Ruegg D, Hohlfeld R, Engel AG (1992) Autoaggressive myocytotoxic T lymphocytes expressing an unusual γ/δ T cell receptor. J Exp Med 176:1785-1789

40. O'Hanlon TP, Messersmith WA, Dalakas MC, Plotz PH, Miller FW (1995) Gamma delta T cell receptor gene expression by muscle-infiltrating lymphocytes in the idiopathic inflammatory myopathies. Clin Exp Immunol 100:519-528

41. Mantegazza R, Bernasconi P, Torchiana E et al (1994) Molecular analysis of T cell receptor repertoire of T cell infiltrates in sporadic and familial inclusion body myositis. Muscle Nerve 17 (Suppl 1):117

42. Lindberg C, Oldfors A, Tarkowski A (1994) Restricted use of T cell receptor V genes in endomysial infiltrates of patients with inflammatory myopathies. Eur J Immunol 24:2659-2663

43. O'Hanlon TP, Dalakas MC, Plotz PH, Miller FW (1994) The $\alpha\beta$ T-cell receptor repertoire in inclusion body myositis: diverse patterns of gene expression by muscle-infiltrating lymphocytes. J Autoimmunity 7:321-333

44. Fyhr IM, Moslemi AR, Mosavi AA, Lindberg C, Tarkowski A, Oldfors A (1997) Oligoclonal expansion of muscle infiltrating T cells in inclusion body myositis. J Neuroimmunol 79:185-189

45. Fyhr IM, Moslemi AR, Lindberg C, Oldfors A (1998) T cell receptor β-chain repertoire in inclusion body myositis. J Neuroimmunol 91:129-134

46. Bender A, Behrens L, Engel AG, Hohlfeld R (1998) T-cell heterogeneity in muscle lesions of inclusion body myositis. J Neuroimmunol 84:86-91

47. Janeway CA, Bottomly K (1994) Signals and signs for lymphocyte responses. Cell 76:275-285

48. Andreetta F, Bernasconi P, Torchiana E, Baggi F, Cornelio F, Mantegazza R (1995) T-cell infiltration in polymyositis is characterized by coexpression of cytotoxic and T-cell-activating cytokine transcripts. Ann N Y Acad Sci 756:418-420

49. Lundberg I, Brengman JM, Engel AG (1995) Analysis of cytokine expression in muscle in inflammatory myopathies, Duchenne dystrophy, and non-weak controls. J Neuroimmunol 63:9-16

50. Lundberg I, Ulfgren AK, Nyberg P, Andersson U, Klareskog L (1997) Cytokine production in muscle tissue of patients with idiopathic inflammatory myopathies. Arthritis Rheum 40:865-874

51. Tews DS, Goebel HH (1996) Cytokine expression profile in idiopathic inflammatory myopathies. J Neuropathol Exp Neurol 55:342-347
52. Authier FJ, Mhiri C, Chazaud B, Christov P, Barlovatz-Meimon G, Gherardi RK (1997) Interleukin-1 expression in inflammatory myopathies: evidence of marked immunoreactivity in sarcoid granulomas and muscle fibres showing ischaemic and regenerative changes. Neuropathol Appl Neurobiol 23:132-140
53. De Bleecker JL, Meire VI, Declercq W, Van Aken EH (1999) Immunolocalization of tumor necrosis factor-alpha and its receptors in inflammatory myopathies. Neuromusc Disord 9:239-246
54. Tateyama M, Nagano I, Yoshioka M, Chida K, Nakamura S, Itoyama Y (1997) Expression of tumor necrosis factor-alpha in muscles in polymyositis. J Neurol Sci 146:45-51
55. Confalonieri P, Bernasconi P, Cornelio F, Mantegazza R (1997) Transforming growth factor-beta1 in polymyositis and dermatomyositis correlates with fibrosis but not with mononuclear cell infiltrate. J Neuropathol Exp Neurol 56:479-484
56. Wahl SM (1994) Transforming growth factor β: the good, the bad and the ugly. J Exp Med 180:1587-1590
57. Gamble JR, Khew-Goodall Y, Vadas MA (1993) Transforming growth factor-beta inhibits E-selectin expression on human endothelial cells. J Immunol 150:4494-4503
58. Hurwitz AA, Lyman WD, Berman JW (1995) Tumor necrosis factor α and transforming growth factor β up-regulate astrocyte expression of monocyte chemoattractant protein-1. J Neuroimmunol 57:193-198
59. Luster AD (1998) Chemokines. Chemotactic cytokines that mediate inflammation. N Engl J Med 338:436-445
60. Carr MW, Roth SJ, Luther E, Rose SS, Springer TA (1994) Monocyte chemoattractant protein 1 acts as a T-lymphocyte chemoattractant. Proc Natl Acad Sci USA 91:3652-3656
61. Allavena P, Bianchi G, Zhou D, van Damme J, Jilek P, Sozzani S, Mantovani A (1994) Induction of natural killer cell migration by monocyte chemotactic protein-1, -2 and -3. Eur J Immunol 24:3233-3236
62. Taub DD, Conlon K, Lloyd AR, Oppenheim JJ, Kelvin DJ (1993) Preferential migration of activated CD4+ and CD8+ T cells in response to MIP-1α and MIP-1β. Science 260:355-358
63. Adams EM, Kirkley J, Eidelman G, Dohlman J, Plotz PH (1997) The predominance of beta (CC) chemokine transcripts in idiopathic inflammatory muscle diseases. Proc Assoc Am Physicians 109:275-285
64. Confalonieri P, Bernasconi P, Megna P, Galbiati S, Cornelio F, Mantegazza R (2000) Increased expression of β-chemokines in muscle of patients with inflammatory myopathies. J Neuropathol Exp Neurol 59:164-169
65. Liprandi A, Bartoli C, Figarella-Branger D, Pellissier J-F, Lepidi H (1999) Local expression of monocyte chemoattractant protein-1 (MCP-1) in idiopathic inflammatory myopathies. Acta Neuropathol 97:642-648
66. De Rossi M, Bernasconi P, Baggi F, de Waal Malefyt R, Mantegazza R (2000) Cytokines and chemokines are both expressed by human myoblasts: possible relevance for the immune pathogenesis of muscle inflammation. Int Immunol 12:1329-1335
67. Sugiura T, Kawaguchi Y, Harigai M et al (2000) Increased CD40 expression on muscle cells of polymyositis and dermatomyositis: role of CD40-CD40 ligand interaction in IL-6, IL-8, IL-15, and monocyte chemoattractant protein-1 production. J Immunol 164:6593-6600

68. Birkedal-Hansen H (1995) Proteolytic remodeling of extracellular matrix. Curr Opin Cell Biol 7:728-735
69. Goetzl EJ, Banda MJ, Leppert D (1996) Matrix metalloproteinases in immunity. J Immunol 156:1-4
70. Choi Y-C, Dalakas MC (2000) Expression of matrix metalloproteinases in the muscle of patients with inflammatory myopathies. Neurology 54:65-71
71. Berke G (1994) The binding and lysis of target cells by cytotoxic lymphocytes: molecular and cellular aspects. Ann Rev Immunol 12:735-773
72. Goebels N, Michaelis D, Engelhardt M et al (1996) Differential expression of perforin in muscle-infiltrating T cells in polymyositis and dermatomyositis. J Clin Invest 97:2905-2910
73. Mantegazza R, Bernasconi P, Confalonieri P, Cornelio F (1997) Inflammatory myopathies and systemic disorders: a review of immunopathogenetic mechanisms and clinical features. J Neurol 244:277-287
74. Orimo S, Koga R, Goto K et al (1994) Immunohistochemical analysis of perforin and granzyme a in inflammatory myopathies. Neuromusc Disord 4:219-226
75. Behrens L, Bender A, Johnson MA, Hohlfeld R (1997) Cytotoxic mechanisms in inflammatory myopathies. Co-expression of Fas and protective Bcl-2 in muscle fibres and inflammatory cells. Brain 120:929-938
76. Fyhr IM, Oldfors A (1998) Upregulation of Fas/Fas ligand in inclusion body myositis. Ann Neurol 43:127-130
77. Schneider C, Gold R, Dalakas MC et al (1996) MHC class I-mediated cytotoxicity does not induce apoptosis in muscle fibers nor in inflammatory T cells: studies in patients with polymyositis, dermatomyositis, and inclusion body myositis. J Neuropathol Exp Neurol 55:1205-1209
78. Spuler S, Engel AG (1998) Unexpected sarcolemmal complement membrane attack complex deposits on nonnecrotic muscle fibers in muscular dystrophies. Neurology 50:41-46
79. Askanas VA, Engel WK , Alvarez RB (1998) Fourteen newly recognized proteins at the human neuromuscular junctions and their nonjunctional accumulation in inclusion-body myositis. Ann N Y Acad Sci 841:28-56
80. Li M, Dalakas MC (2000) The muscle mitogen-activated protein kinase is altered in sporadic inclusion body myositis. Neurology 54:1665-1669
81. Medsger TA, Dawson WN, Masi AT (1970) The epidemiology of polymyositis. Am J Med 48:715-723
82. Amato AA, Barohn RJ (2000) Evaluation and treatment of the idiopathic inflammatory myopathies. Neurologist 6:267-287
83. Oldfors A, Lindberg C (1999) Inclusion body myositis. Curr Opin Neurol 12:527-533
84. Koffman BM, Rugiero M, Dalakas MC (1998) Immune-mediated conditions and antibodies associated with sporadic inclusion body myositis. Muscle Nerve 21:115-117
85. Hengstman GJ, van Engelen BG, Badrising UA, van de Hoogen FH, van Venrooij WJ (1998) Presence of the anti-Jo-1 autoantibody excludes inclusion body myositis. Ann Neurol 44:423
86. Callen JP (2000) Dermatomyositis. Lancet 355:53-57
87. Mastaglia FL (2000) Treatment of autoimmune inflammatory myopathies. Curr Opin Neurol 13:507-509
88. Mantegazza R, Antozzi C, Cornelio F, Di Donato S (2001) Clinical trials in muscle disorders. In: Biller J, Bogousslavsky J (eds) Clinical trials in neurologic practice. Butterworth-Heinemann, Woburn, pp 311-325

89. Griggs RC, Rose MR (1998) Evaluation of treatment for sporadic inclusion-body myositis. In: Askanas V, Serratrice G, Engel WK (eds) Inclusion-body myositis and myopathies. Cambridge University Press, Cambridge, pp 331-350

90. Dalakas MC, Koffman B, Fujii M, Spector S, Sivakumar K, Cupler E (2001) A controlled study of intravenous immunoglobulin combined with prednisone in the treatment of IBM. Neurology 56:323-327

Chapter 7

Immunological Mechanisms of Paraneoplastic Nervous System Diseases

B. Giometto, P. Nicolao, T. Scaravilli, M. Vianello, B. Vitaliani,
A.M. Ferrarini, B.Tavolato

Introduction

The term "paraneoplastic neurological diseases" (PND) (also known as "remote neurological effect of cancer") encompasses a wide group of disorders associated with systemic malignancies which, in order to fulfil the criteria for the definition of this disorder, must not directly invade, compress or metastasise to the nervous system.

In their seminal publication, Henson and Urich [1] described several clinico-pathological entities associated with tumours:

- *Encephalomyelitis.* The entity that presently goes under the name of "encephalomyelitis" was first reported by Greenfield [2], who noted perivascular infiltration and inflammatory nodules in the brain stem and spinal cord of a patient with bronchial carcinoma. In 1954, Henson et al [3] made the observation that the term "subacute cerebellar degeneration" seemed inappropriate in some of their cases, in view of the well-defined inflammatory process they described in the subthalamic nucleus and spinal cord. It was suggested that, because of the severity of the inflammatory changes, the pathology they described was indistinguishable from viral encephalitis. Eventually, Henson et al [4] coined the term "encephalitis with carcinoma", which includes limbic and brain stem encephalitis and the sensory neuropathy described by Denny-Brown in 1948 [5], thus emphasising the frequent global involvement of both central and peripheral nervous system in these syndromes. Henson and Urich found 67 case reports in the literature that, on the basis of the predominant localisation of the lesions, could be subdivided into a) limbic, b) bulbar, c) cerebellar, d) myelitis and e) ganglio-radiculo-neuritis or sensory neuropathy.
- *Cortical cerebellar degeneration.* Patients presenting with this syndrome were divided into two groups: 27 of them presented with a purely degenerative process confined to the cerebellum, whereas in 17 there was, in addition to degeneration, a prominent inflammatory reaction.
- *Peripheral neuropathy.* This group included sensory, sensorimotor, remitting and relapsing, progressive and mild terminal neuropathies. Guillain-Barré

Department of Neurologic and Psychiatric Sciences, Second Neurological Clinic, University of Padua, Via Vendramini 7, 35137 Padua, Italy. e-mail: giometto@ux1.unipd.it syndrome was also included by Henson and Urich, although they consider this a rare complication of cancer.

– *Muscular and neuromuscular disorders*. These were further classified as polymyopathy, disorders of neuromuscular transmission, dermatomyositis and polymyositis and endocrine-metabolic disorders. Lambert-Eaton syndrome is also included in this group.

In addition, there is a miscellaneous group which includes disorders that are sometimes related to neoplasms, such as opsoclonus-myoclonus and optic neuritis, and the retinopathies which can be neoplastic. Table 1 lists most of the known syndromes.

Anti-Neuronal Antibodies

PND are rare disorders and the final diagnosis relies on neuropathological examination. An important breakthrough, therefore, was obtained in the eighties [6] by

Table 1. Paraneoplastic neurological syndromes

Central nervous system
    Encephalomyelitis
        Limbic encephalitis
        Brain stem encephalitis
        Cerebellar degeneration
    Opscoclonus-myoclonus
    Pure cerebellar degeneration
    Paraneoplastic visual syndrome
    Stiff-person syndrome
    Necrotising myelopathy
    Myelitis
    Motor neuron syndrome
    Subacute motor neuronopathy

Peripheral nervous system and dorsal root ganglia
    Sensory neuronopathy
    Autonomic neuropathy
    Acute sensorimotor neuropathy
    Chronic sensorimotor neuropathy
    Vasculitic neuropathy
    Neuromyotonia

Neuromuscular junction and muscle
    Lambert-Eaton myasthenic syndrome
    Myasthenia gravis
    Polymyositis/Dermatomyositis
    Acute necrotising myopathy
    Cachectic myopathy
    Myotonia

Italics indicate disorders associated with autoantibodies

the demonstration of specific autoantibodies, in both serum and cerebrospinal fluid, reacting with tumours and with antigens in the central nervous system (CNS) [7]. Interestingly, detection of these antibodies supports the hypothesis, already suggested by Russell Brain in 1951 [8], that most PND might be the result of an immunological response triggered by a tumour antigen that cross-reacts with a protein expressed by the nervous system. The detection of these autoantibodies has become a useful tool both for the diagnosis of neurological disorders and for directing the search for an underlying tumour.

The first report of antineuronal antibodies in patients with paraneoplastic disorders appeared in 1965, when Wilkinson and Zeromski [9] found them in the serum of four individuals suffering from small-cell lung cancer and sensory neuropathy. Furthermore, Trotter et al [10] were able to detect antibodies reacting with human Purkinje cells in the serum of a 21-year old woman with Hodgkin's lymphoma, whilst Greenlee and Brashear [11] described two patients with ovarian cancer and high titres of anti-Purkinje cell antibodies.

The concept that antibodies were present in the serum of patients suffering from various types of tumours gave new impetus to the study of paraneoplastic disorders. Moreover, it became clear that some tumours were preferentially associated with a particular neurological disorder. Purkinje cell degeneration is most commonly found in association with gynaecological and lung neoplasms; opsoclonus-myoclonus is particularly frequent in children with neuroblastoma, whereas the Lambert-Eaton myasthenic syndrome occurs in patients with small-cell lung cancer.

Table 2 classifies PND according to the specific antibody detected, the neurological syndrome and the most frequently associated tumour.

Anti-Yo, reacting with the cytoplasm of Purkinje cells, was found almost exclusively in women with Purkinje cell degeneration, although a few men with this antibody have been found (see Chap. 8). The neural proteins of 34 and 62 kDa, with which this antibody reacts were cloned from the gene in 1990 by Sakai et al [12]. The gene products are called cerebellar degeneration-related proteins, CDR1 (CDR-34) and CDR2 (CDR-62) [13]. Unlike anti-Hu antibodies, which are expressed in all small-cell lung cancers, the CDRs are expressed only in tumours of patients with anti-Yo-positive PND [14, 15].

Anti-Hu antibodies were detected in a disorder referred to as "paraneoplastic encephalomyelopathy and sensory neuronopathy". They are almost always associated with small-cell lung cancer. Their related antigens are found in neuronal nuclei and on immunohistochemistry the neuronal nuclei are specifically stained by the patient sera sparing the nucleolus (Fig. 1A, B). At least four Hu antigens have been cloned and isolated: HuD [16], HuC [17], He l-N1 and He l-N2 [18, 19, 20]. These proteins are all expressed by small-cell lung cancer cells [19, 20, 21] and are members of a family of human neuronal RNA-binding proteins homologous to the *Drosophila* embryonic lethal abnormal visual protein (ELAV), which is involved in the normal development of the fly nervous system [16, 22]. They contain three RNA-binding domains that encode proteins that seem to play an important role in the control of mRNA [23] and maintenance of the neuronal

Table 2. Classification of PND according to the specific antibody detected, the neurological syndrome and the most frequently associated tumor

Antibody	Paraneoplastic neurological syndrome	Most frequently associated cancers	Immunohisto-chemistry	Western immunoblot	Antigen
Anti-Hu	Encephalomyelitis, sensory neuropathy	SCLC, non-SCLC	Strong staining of all neuronal nuclei and weaker staining of cytoplasm	38- to 40-kDa reactive triplet bands on extracts of crude or isolated CNS neurons/nuclei and SCLC. Recombinant protein 43kD (HuD)	Neuron specific RNA/DNA binding protein, role in RNA processing of neurons
Anti-Yo	Acute to subacute pancerebellar syndrome, dysarthria and upbeat nystagmus	Ovarian, breast, uterus	Purkinje-cell cytoplasm and axons, coarse granular staining	34 and 62 kD band from Purkinje cell extracts and breast and ovary tumors of patients, reactive with recombinant CDR 34 and CDR 62	DNA binding, gene transcription regulators, carry leucine zipper and zinc finger motifs
Anti-Tr	More slowly developing cerebellar syndrome, less frequent dysarthria and nystagmus	Hodgkin Disease	Purkinje cell cytoplasm, fine granular staining in the molecular layer	No reactive protein detected as yet	n.k.
Anti-Ri	Opsoclonus-myoclonus	Breast, SCLC	All nuclei of CNS	55- and 80-kDa neuronal protein. 55 kD recombinant protein	RNA-binding protein, regulates metabolism in a subset of developing neurons
Anti-Amphiphysin	Stiff-person syndrome	Breast	Synapses of CNS neurons	128 kD protein on neuronal extracts	Amphiphysin synaptic vesicle protein, function unknown
anti-VGCC	LEMS	SCLC	nK	nK	VGCC
anti-Ta	Limbic encephalitis	Testicular tumors	Nucleus, perikarion	40 Kd	Unknown function

SCLC, Small-cell lung carcinoma; *VGCC*, voltage gated calcium channels; *LEMS*, Lambert-Eaton myasthenic syndrome; *n.k.*, not known phenotype [24]. The antibodies are directed against two of the RNA-binding domains of HuD and cross-react with both HuC and He l-N1.

Anti-Ri is the rarest of this group: it has the staining properties of anti-Hu, but is limited to neurons of the CNS. It was demonstrated in adult patients with opsoclonus-myoclonus and breast cancer. In Western blot of neuronal extracts the antibody shows two separate bands of 55 and 80 kDa [25]. The sequence that encodes the neuronal protein recognised by anti-Ri has been cloned and the recombinant protein has been called Nova [26].

Anti-P/Q-type voltage-gated calcium channel (VGCC) antibodies are found in patients with the Lambert-Eaton myasthenic syndrome (see Chap. 3), but can also be associated with cerebellar ataxia (see Chap. 8). Antibodies to amphiphysin (Fig. 1C), and the recently described Ma1 (Fig. 1D) and Ma2 antigens, appear to be much less common, but may be helpful in predicting the presence of breast and testicular tumours respectively.

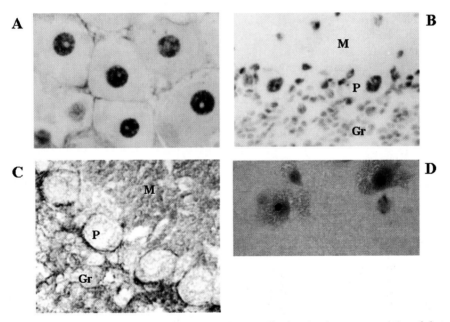

Fig. 1A-D. × 400. **A** Anti-Hu: dorsal root ganglionic cells showing intense reactivity of the neuronal nuclei. **B** Anti Hu: cerebellar cortex with nuclear staining of the Purkinje cell (*P*), granular neurons (*Gr*) and no reactivity in the molecular layer (*M*). **C** Anti-amphiphysin: staining of the neuropil of the molecular (*M*) and granular (*Gr*) layer of cerebellum, sparing the nucleus and cytoplasm of Purkinje (*P*) and granular neurons. **D** Anti-MA1: cortical neurons showing reactivity of the nucleolus. (Courtesy of Dr. Dalmau)

Immunological Mechanisms

The identification of the antibodies and the finding that they are present in the cerebrospinal fluid and may be synthesised intrathecally appears to settle the issue about whether paraneoplastic syndromes are immunological disorders. In particular, the autoimmune disorder seems to be triggered by the expression in tumours of proteins that are normally expressed by neurons only; these can be called onconeuronal antigens.

In this model, the immune system recognises the neuronal proteins as *foreign* when ectopically expressed in tumour cells; their expression in these cells leads to an immune response which on the one hand suppresses tumour growth, but on the other leads to the destruction of neurons. However, to date, firm evidence in favour of autoimmunity in paraneoplastic disorders exists only in the Lambert-Eaton myasthenic syndrome [27], in which passive transfer and in vitro experiments have confirmed their pathogenetic role (see Chap. 3). The antibodies developed in these patients are against VGCC situated in the presynaptic active zones of the neuromuscular junction and interfere with the quantal release of acetylcholine from nerve terminals [28]. Small-cell lung cancer, the tumour most commonly involved in Lambert-Eaton myasthenic syndrome, also expresses calcium channel antigens identified by these antibodies and a cross-reactive mechanism of damage has been demonstrated [29].

A similar autoimmune hypothesis has been postulated, but not proven, for other paraneoplastic syndromes involving the CNS. In these cases, the expression of onconeural antigens in patients' tumours is more selective and appears to be directly related to the presence of the PND [15, 21]. However, the role of the immune response against these antigens in causing the neurological disorder is still unclear. Most of these antigens are localised in the cytoplasm or in the nuclei, and it is difficult to understand how antibodies could target the neurons. Although there is evidence that neurons may non-specifically take up antibodies, experiments have failed to reproduce the disease by injection of antibodies into experimental animals. In these PND, several mechanisms have been postulated and both autoantibodies and T cells have been implicated.

Antibody-Related Mechanisms

Anti-Hu

Hu antigens are expressed in cell nuclei and not on the cell surface; this suggests that they are unavailable as targets for the immune system. In this respect, neither passive transfer of anti-Hu antibodies nor direct immunisation with the Hu antigen has reproduced the neurological disease in mice despite the fact that immunised animals developed high titres of anti-Hu antibodies [30]. These findings were confirmed in vitro [31]. The only clue that Hu antibodies may have a direct pathogenetic role comes from the finding that anti-Hu antibodies destroy rat cerebellar granule cells in vitro [32]. However, anti-Hu-positive sera were shown to be toxic to tumour cell lines even if the cells did not express Hu antigens, suggesting that the possible effect on cells might not be Hu-antigen-mediated [33]. Deposits of IgG have been found in the neurons of autopsy cases of paraneoplastic encephalomyelitis and sensory neuropathy [34, 35], which tends to support an antibody-mediated pathology. However, deposits of IgG in the neurons could have occurred after the death of the patient; in animals with circulating anti-Hu antibodies, intraneuronal IgG may be observed if the animal is sacrificed but the brain is not immediately perfused with saline and paraformaldehyde. In animals perfused immediately after death, deposits of intraneuronal IgG were not observed [36, 30].

Anti-Yo

Pathological studies reported a total absence of Purkinje cells, with only mild inflammatory cell infiltration in the cerebellum of patients with paraneoplastic cerebellar degeneration and anti-Yo antibodies [37, 38]. A neurotoxic mechanism of damage by the antibody interfering with a leucin-zipper protein has been suggested [13]. However, attempts to reproduce an animal model of paraneoplastic cerebellar degeneration by passive transfer of anti-Yo antibodies or active immunisation with the Yo antigen have failed, as with anti-Hu. The injection of human anti-Yo in Lewis rats after disruption of the blood-brain barrier by inducing experimental allergic encephalomyelitis did not reproduce the disease [39]. Although internalisation of human antibodies in the Purkinje cells was demonstrated, the animals were neurologically normal and no pathological changes were detectable in the cerebellum [39]. Graus et al [40] have shown that anti-Yo injected into the cerebrospinal fluid is taken up by guinea pig Purkinje cells without causing cell damage. Other authors immunised with Yo recombinant protein four different strains of mice (expressing different MHC molecules); all animals produced a high anti-Yo antibody titre without evidence of neurological disease [41]. The passive transfer of murine mononuclear cells activated with anti-Yo in severe combined immunodeficiency mice was similarly ineffective in inducing the disease [42].

T-Cell Related Mechanisms

The observation that the presence of PND autoantibodies has failed to produce the disease in experimental animals implies a possible role of alternative immune mechanisms (e.g., cytotoxic T lymphocytes or other cytolytic killer cell activity). Several studies have suggested that T cells play a role in the PND.

T cells can be detected in pathological brain specimens of patients with PND as dense perivascular infiltrates [43]. Extensive infiltrates predominate in the nervous system of anti-Hu, anti-Ri and anti-Ma antibody-producing patients, while milder mononuclear cell infiltrates have been detected in patients with anti-Yo-associated cerebellar degeneration [43]. The lymphocytes express the CD4 and CD8 phenotypes, and the CD8 lymphocytes can also be detected in the brain parenchyma in close contact with neurons.

The hypothesis that a cytotoxic T-cell-mediated mechanism is involved in some PND is also supported by a study of T-cell receptor usage in inflammatory infiltrates both of the CNS and in the tumours of patients with anti-Hu syndrome. In this study all five patients showed a limited Vβ T-cell receptor repertoire [44], suggesting that T lymphocytes are specifically recruited and targeted to neuronal and tumour antigens. High expression of MHC class I and II antigens was demonstrated in the majority of small-cell lung cancer tumours from patients with anti-Hu syndrome, suggesting that the ability of the tumours to present Hu antigens to the immune system could be critical in triggering the immune response leading to the paraneoplastic disorder [45].

Finally, a large percentage of activated T cells have recently been demonstrated in the CSF of patients with acute or chronic PCD; in addition, the CD4 cells revealed a Th1 cytokine profile (IFN-γ, TNF-α and IL-2), supporting the hypothesis that an active cell-mediated autoimmune response is ongoing within the CNS [46]. In addition, a recent report showed an expanded population of class I restricted CD8 cytotoxic T lymphocytes specific for the CDR2 antigen in the blood of patients with paraneoplastic cerebellar degeneration [47]. This was the first demonstration that onconeural antigen-specific cytotoxic T lymphocytes are present in the peripheral blood of patients with paraneoplastic disorders and could mediate effective tumour suppression whilst inducing the paraneoplastic disorder. Moreover, these cells have the potential to recognise intracellular onconeural antigens processed and presented via MHC-I molecules.

The only limitation of this model is the reported lack of constitutive expression of MHC-I antigen by neuronal cells. However, it has been demonstrated that MHC molecules may be inducible in neurons in vitro [48]. A recent study using immunohistochemistry showed that a subset of neurons (the mature hippocampus and Purkinje cells) normally express MHC-I antigens. Indeed, it has been postulated that MHC-I molecules may play a non-classical and novel role in neuronal signalling and activity-dependent changes in synaptic connectivity [49]. Although not all components of the MHC-I signalling system have been studied in neurons, the data reported above suggest that selected neurons may be particularly susceptible to T-cell mediated autoimmune attack.

Conclusions

Although no animal models of PND with CNS involvement have been established so far, and consequently the mechanisms of damage are still the subject of controversy, the study of these rare diseases could provide insights into basic neuroimmunology, neurobiology and tumour immunology [50]. PND provide an interesting framework within which to analyse the relationship between neurons and systemic immune responses directed, in this case, against tumour cells. In addition, the characterisation of onconeuronal antigens has opened an interesting window into basic neurobiology; the antigens expressed by tumour cells of most patients with PND are nuclear proteins involved in the differentiation and maturation of neuronal cells. Finally, the recognition that CDR2-specific cytotoxic T lymphocytes, present in the blood of patients with PCD and gynaecological cancers, could mediate effective tumour suppression, provide oncologists with a model for a cancer immunotherapy.

References

1. Henson RA, Urich H (1982) Cancer and the Nervous System. Blackwell Scientific, Oxford
2. Greenfield JG (1934) Subacute spinocerebellar degeneration occurring in elderly patients. Brain 57:161-176
3. Henson RA, Russell DS, Wilkinson M (1954) Carcinomatous neuropathy and myopathy. A clinical and pathological study. Brain 77:82-121
4. Henson RA, Hoffman HL, Urich H (1965) Encephalomyelitis with carcinoma. Brain 88:449-464
5. Denny-Brown D (1948) Primary sensory neuropathy with muscular changes associated with carcinoma. J Neurol Neurosurg Psychiat 11:73-87
6. Posner JB (1995) Paraneoplastic syndromes. In: Posner JB (ed) Neurologic complication of cancer. Davis, Philadelphia, pp 353-383. (Contemporary Neurology series 45)
7. Giometto B, Tavolato B, Graus F (1999) Autoimmunity in paraneoplastic neurological disorders. Brain Pathol 9:261-273
8. Russell Brain W, Daniel PM, Greenfield JG (1951) Subacute cortical cerebellar degeneration and its relation to carcinoma. J Neurol Neurosurg Psychiatry 14:59-75
9. Wilkinson PC, Zeromski J (1965) Immunofluorescent detection of antibodies against neurons in sensory carcinomatous neuropathy. Brain 88:529-538
10. Trotter JL, Hendin BA, Osterland CK (1976) Cerebellar degeneration with Hodgkin's disease. Arch Neurol 33:660-661
11. Greenlee JE, Brashear HR (1983) Antibodies to cerebellar Purkinje cells in patients with paraneoplastic cerebellar degeneration and ovarian carcinoma. Ann Neurol 14:609-613
12. Sakai K, Mitchell DJ, Tsukamoto T, Steinman L (1990) Isolation of a cDNA clone encoding an autoantigen recognised by an anti-neuronal cell antibody from a patient with paraneoplastic cerebellar degeneration (cloning of the PCD-17 autoantigen). Ann Neurol 28:692-698
13. Fatallah-Shaykh H, Wolf S, Wong E et al (1991) Cloning of a leucine-zipper protein recognized by the sera of patients with antibody-associated paraneoplastic cerebellar degeneration. Proc Natl Acad Sci USA 88:3451-3545
14. Sakai K, Negami T, Yoshioka A, Hirose G (1992) The expression of a cerebellar degeneration-associated neural antigen in human tumor line cells. Neurology 42:361-366
15. Furneaux HM, Rosenblum MK, Dalmau J et al (1990) Selective expression of Purkinje-cell antigens in tumor tissue from patients with paraneoplastic cerebellar degeneration. N Engl J Med 227:1844-1851
16. Szabo A, Dalmau J, Manley G et al (1991) HuD a paraneoplastic encephalomyelitis antigen, contains RNA-binding domains and is homologous to Elav and Sex-lethal. Cell 67:325-333
17. Sakai K, Gofuku M, Kitagawa Y et al (1994) A hippocampal protein associated with paraneoplastic neurologic syndrome and small cell lung carcinoma. Bioch Biophys Res Commun 199:1200-1208

18. King PH (1994) Hel-N2: a novel isoform of Hel-N1 which is conserved in rat neural tissue and produced in early embryogenesis. Gene 151:261-265
19. King PH, Dropcho EJ (1996) Expression of Hel-N1 and Hel-N2 in small-cell lung carcinoma. Ann Neurol 39:679-681
20. Sekido Y, Bader SA, Carbone DP et al (1994) Molecular analysis of the HuD gene encoding a paraneoplastic encephalomyelitis antigen in human lung cancer cell lines. Cancer Res 54:4988-4992
21. Dalmau J, Furneaux HM, Cordon-Cardo C, Posner JB (1992) The expression of the Hu (paraneoplastic encephalomyelitis/sensory neuronopathy) antigen in normal and tumor tissues. Am J Pathol 141:881-886
22. King PH, Levine TD, Fremeau RT Jr, Keene JD (1994) Mammalian homologs of Drosophila ELAV localized to a neuronal subset can bind in vitro to the 3° UTR of mRNA encoding the Id transcriptional repressor. J Neurosci 14:1943-1952
23. Liu J, Dalmau J, Szabo A et al (1995) Paraneoplastic encephalomyelitis antigens bind to the AU-rich elements of mRNA. Neurology 45:544-550
24. Marusich MF, Weston JA (1992) Identification of early neurogenic cells in the normal crest lineage. Dev Biol 149:295-306
25. Luque FA, Furneaux HM, Ferziger R et al (1991) Anti-Ri: an antibody associated with paraneoplastic opsoclonus and breast cancer. Ann Neurol 29:241-251
26. Buckanovich RJ, Posner JB, Darnell RB (1993) Nova, the paraneoplastic Ri antigen, is homologous to an RNA-binding protein and is specifically expressed in the developing motor system. Neuron 11:657-672
27. Lang B, Newsom-Davies J, Prior C, Wray D (1983) Autoimmune aetiology for myasthenic (Eaton Lambert) syndrome. Lancet ii:224-226
28. Leys K, Lang B, Johnston I, Newsom-Davies J (1991) Calcium channel autoantibodies in the Lambert-Eaton myastenic syndrome. Ann Neurol 29:307-314
29. Lennon VA (1996) Calcium channel and related paraneoplastic disease autoantibodies. In: Peter JB, Shoenfeld Y (eds) Autoantibodies. Elsevier Science, Amsterdam, pp 139-147
30. Sillevis-Smitt PAE, Manley GT, Posner JB (1994) High titer antibodies but no disease in mice immunized with the paraneoplastic antigen HuD. Neurology 43:2049-2054
31. Hormigo A, Lieberman F (1994) Nuclear localization of anti-Hu antibody is not associated with in vitro cytotoxicity. J Neuroimmunol 55:205-212
32. Greenlee JE, Parks TN, Jaeckle KA (1993) Type Ia ("anti-Hu") antineuronal antibodies produce destruction of rat cerebellar granule neurons in vitro. Neurology 43:2049-2054
33. Verschuuren JJ, Dalmau J, Hoard R, Posner JB (1997) Paraneoplastic anti-Hu serum: studies on human tumor cell lines. J Neuroimmunol 79:202-205
34. Dalmau J, Graus F, Rosenblum MK, Posner JB (1991) Detection of the anti-Hu antibody in specific regions of the nervous system and tumor from patients with paraneoplastic encephalomyelitis/sensory neuropathy. Neurology 41:1757-1764
35. Graus F, Elkon KB, Cordon-Cardo C, Posner JB (1986) Sensory neuronopathy and small cell lung cancer. Antineuronal antibody that also reacts with the tumor. Am J Med 80:45-52
36. René R, Ferrer I, Graus F (1996) Clinical and immunohistochemical comparison of in vivo injected anti-Hu and control IgG in the nervous system of the mouse. Eur J Neurol 3:319-323
37. Giometto B, Marchiori GC, Nicolao P et al (1997) Subacute cerebellar degeneration with anti-Yo autoantibodies: immunohistochemical analysis of the immune reaction in the central nervous system. Neuropathol Appl Neurobiol 23:468-474

38. Verschuuren JJ, Chuang L, Rosenblum MK et al (1996) Inflammatory infiltrates and complete absence of Purkinje cells in anti-Yo-associated paraneoplastic cerebellar degeneration. Acta Neuropathol (Berl) 91:519-525

39. Greenlee JE, Burns JB, Rose JW et al (1995) Uptake of systemically administered human cerebellar antibody by rat Purkinje cells following blood-brain barrier disruption. Acta Neuropathol 89:341-345

40. Graus F, Illa I, Agusti M et al (1991) Effect of intraventricular injection of an anti-Purkinje cell antibody (anti-Yo) in a guinea pig model. J Neurol Sci 106:82-87

41. Tanaka M, Tanaka K, Onodera O, Tsuji S (1995) Trial to establish an animal model of paraneoplastic cerebellar degeneration with anti-Yo antibody. Mouse strains bearing different MHC molecules produce antibodies on immunization with recombinant Yo protein, but do not cause Purkinje cell loss. Clin Neurol Neurosurg 97:95-100

42. Tanaka K, Tanaka K, Igarashi S et al (1995) Trial to establish an animal model of paraneoplastic cerebellar degeneration with anti-Yo antibody. Passive transfer of murine mononuclear cells activated with recombinant Yo protein to paraneoplastic cerebellar degeneration lymphocytes in severe combined immunodeficiency mice. Clin Neurol Neurosurg 97:101-105

43. Scaravilli F, An SF, Groves M, Thom M (1999) The neuropathology of paraneoplastic syndromes. Brain Pathol 9:251-260

44. Voltz R, Dalmau J, Posner JB, Rosenfeldt MR (1998) T-cell receptor analysis in anti-Hu associated paraneoplastic encephalomyelitis. Neurology 51:1146-1150

45. Dalmau J, Graus F, Cheung NK et al (1995) Major histocompatibilility proteins, anti-Hu antibodies and paraneoplastic encephalolomyelitis in neuroblastoma and small cell lung cancer. Cancer 75:99-109

46. Albert ML, Austin LM, Darnell RB (2000) Detection and treatment of activated T cells in the cerebrospinal fluid of patients with paraneoplastic cerebellar degeneration. Ann Neurol 47:9-17

47. Albert ML, Darnell JC, Bender A et al (1998) Tumor-specific killer cells in paraneoplastic cerebellar degeneration. Nature Med 4:1321-1324

48. Neumann H, Cavalié A, Jenne DE, Wekerle H (1995) Induction of MHC class I genes in neurons. Science 269:549-552

49. Corriveaux RA, Huh GS, Shatz CJ (1998) Regulation of class I MHC gene expression in the developing and mature CNS of neural activity. Neuron 21:505-520

50. Darnell RB (1996) Onconeural antigens and the paraneoplastic neurologic disorders: at the intersection of cancer, immunity, and the brain. Proc Natl Acad Sci USA 93:4529-4536

Chapter 8

Clinical Presentation and Mechanisms of Immune-Mediated Cerebellar Ataxia

J. HONNORAT

Introduction

The brain has traditionally been regarded as immunologically privileged because of the existence of the blood-brain barrier, the absence of conventional lymphatic drainage and the unusual tolerance of the brain to transplanted tissue. However, over the last decades, clinical evidence has accumulated indicating that the immune system may play an important role in some central nervous system diseases usually regarded as degenerative. The best-known example is paraneoplastic cerebellar ataxia (PCA), which is thought to involve autoimmune cross-reaction between tumour and nervous system antigens. In the past 15 years, several antibodies directed against neuronal and tumoral antigens have been described in association with PCA, leading to the definition of different subtypes of PCA based on their associated antibodies, the clinical evolution and the type of tumour. Circulating antibodies have also been described in patients with non-paraneoplastic cerebellar ataxia (N-PCA), suggesting that the immune system may be involved in certain cases of sporadic cerebellar ataxia. In this review, the clinical presentation of the different subtypes of potentially immune-mediated PCA and N-PCA will be described, and the experimental approaches that have been developed in order to understand the pathogenic role of the immune system in these ataxias will be discussed.

Paraneoplastic Cerebellar Ataxia

PCA is, together with ganglioneuronopathy, one of the most common paraneoplastic neurological diseases (PND) [1]. The first case was probably described by Brouwer in 1919 in a 60-year-old woman with sarcoma of the pelvis [2], but a relationship between cerebellar ataxia and cancer was not widely recognised until 1951 when Brain and colleagues [3] reported four cases of PCA. The syndrome was finally clearly established in 1982 with a review of 50 patients with autopsy-verified PCA [4], and many other patients have since been reported.

Neurologie B, Hôpital Neurologique, 59 Bd Pinel, BP Lyon Montchat, 69394 Lyon Cedex 03, France. e-mail: jerome.honnorat@chu-lyon.fr

The pathological hallmark of PCA is the presence of a marked loss of Purkinje cells and perivascular inflammatory infiltrates, both of which affect all parts of the cerebellum [5, 6]. Less striking changes in the cerebellar cortex include thinning of the molecular layer with variable microglial proliferation and astrocytic gliosis, proliferation of Bergmann astrocytes, and slight thinning of the granular layer with a reduction in the number of granule cells.

The hypothesis that PND and PCA have an autoimmune basis was probably first proposed in 1961 by Russell [7], who suggested that an antigen-antibody reaction might explain the perivascular inflammatory infiltrates seen in the brain, leading several groups to screen serum and cerebrospinal fluid (CSF) from PND patients for autoantibodies. In 1976, Trotter identified a Purkinje-cell-reactive antibody in the serum of a PCA patient [8]. Subsequently, various other antibodies were reported in association with PND [9]. Using clinical and immunological criteria, different PCA subtypes can be defined (Table 1). In some types, cerebellar ataxia occurs in isolation, whereas in others it is only one clinical feature in the setting of extensive nervous system disease. Although much of the data argues in favour of an immune origin, there have only been occasional reports of partial or near-complete remission of PCA following treatment of the primary tumour or the use of corticosteroids or immunoglobulins [10].

PCA with Anti-Yo Antibodies

Anti-Yo antibodies (Yo-Ab) were first described in 1983 in two PCA patients [11]. As shown by indirect immunohistochemical studies of human brain cerebellum,

Table 1. Main subtypes of PCA associated with antibodies

Antibodies present	Clinical presentation	Associated tumor
Anti-Yo	Pure cerebellar ataxia Mainly female patients Average age 61 years	Ovary Breast
Anti-Hu	Frequently associated with PEM/SSN No sex predilection Average age 66 years	Small-cell lung carcinoma
Anti-CV2	Frequently associated with uveitis and neuropathy Mainly male patients Average age 62 years	Small-cell lung carcinoma Thymoma
Anti-Tr, anti-mGluR1	Pure cerebellar ataxia Mainly male patients Average age 30 years	Hodgkin's disease
Anti-Ri	Associated with opsomyoclonus No sex predilection	Breast Lung carcinoma

PEM/SSN, paraneoplastic encephalomyelitis/sensoryneuronopathy these antibodies recognise an antigen expressed with a characteristic coarsely granular pattern specifically in the cytoplasm of cerebellar Purkinje cells (Fig. 1). On immunoblots, Yo-Ab recognise two different 34- and 62-kDa proteins [12].

The clinical features of patients with Yo-Ab are now well known [13, 14]. Most of these patients are female and their average age is 61 years; only two male patients have been described [15, 16]. Cerebellar ataxia precedes the detection of the associated tumour in more than 50% of patients, occasionally by as much as 3 years. Cerebellar signs usually begin with gait ataxia and, over a few weeks or months, progress to severe, usually symmetrical truncal and appendicular ataxia, with dysarthria and often nystagmus (especially rotatory or vertical) [13]. Vertigo is common, and many patients complain of diplopia. Occasionally the onset is more rapid, within a few hours or days [17], or it may be slower. The cerebellar deficit usually stabilises, but by then the patient is often severely incapacitated; most become bed-bound in the first 3 months after diagnosis. Thirty percent of patients die as a result of the debilitating neurological condition [14]. Signs or symptoms of involvement at other levels of the neuraxis are common but usually mild; these include emotional lability, memory deficit, hyporeflexia and mild sensory complaints suggesting peripheral neuropathy [13]. The CSF may be normal, but more typically shows mild pleocytosis, elevated protein levels, an increased IgG concentration and oligoclonal bands. Cranial CT or MRI scans are initially

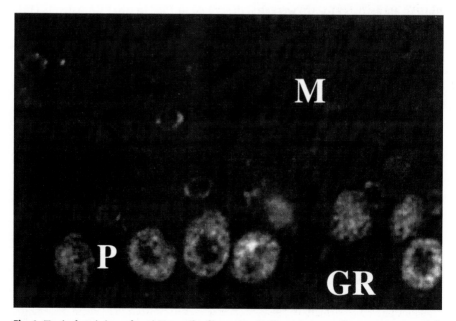

Fig. 1. Typical staining of anti-Yo antibodies using indirect immunofluorescence on adult rat cerebellum. There is intense and granular labeling of Purkinje cell cytoplasms (*P*) with nuclear sparing. Cytoplasm of some neurons are also labeled in the molecular layer (*M*). The granular layer (*GR*) is clear. (Original magnification x 400)

normal, but at a later date show signs of progressive cerebellar atrophy. Usually, the associated tumour is gynaecological [13, 14], the most frequent types being ovary and breast cancer [14]. Tumour progression is the cause of death in more than 50% of patients. Prognosis is worse in patients with ovarian carcinoma (median survival 22 months) and better in patients with breast cancer (median survival 100 months) [14].

There are several arguments in favour of involvement of the immune system in the pathogenesis of PCA with Yo-Ab. A high Yo-Ab titre is seen in the CSF of patients early in the course of the disease [13], and the antibodies are actively synthesised within the CSF compartment [18]. Infiltration with T lymphocytes, especially CD8+ lymphocytes, is observed in some pathological specimens of cerebellum from patients with PCA and Yo-Ab [6, 19]. Yo-Ab have been used to clone three cDNAs coding for cdr1, a 34-kDa antigen located on the long arm of chromosome X near the *FRM1* gene [20]; cdr2, also known as pCD17, a 62-kDa cytoplasmic protein harbouring a helix-leucine zipper motif [21, 22]; and cdr3, also known as CDR62-2 (GenBank accession number L02867).

Cdr2 seems to be the main antigen for Yo-Ab and is widely expressed in breast and ovarian tumours [23], suggesting that it may trigger the immune response in the patient's tumour. Cdr2 interacts specifically through its helical leucine zipper motif with the oncogene *c-Myc* and may down-regulate *c-Myc* function by a mechanism involving sequestration of *c-Myc* in the cytoplasm [24]. All Yo-Ab target the cdr2 helical leucine zipper motif [25] and can prevent the *in vitro* binding of cdr2 to *c-Myc* [24].

Yo-Ab could contribute to Purkinje cell degeneration by disruption of the cdr2/c-Myc interaction in the cytoplasm, leading to aberrant nuclear c-Myc activity and Purkinje cell apoptosis [24]. Neurons, especially Purkinje cell, can efficiently take up IgG [26, 27], which strengthens the possibility that Yo-Ab may be involved in PCA pathogenesis. However, attempts at reproducing the disease in animals have failed. In rodents, prolonged administration of Yo-Ab directly into the CSF via a ventricular catheter [28, 29] or by intraperitoneal administration following blood-brain barrier disruption [30], led to incorporation of antibodies into the Purkinje cell, but did not result in cerebellar injury. Immunisation of animals with cdr2 recombinant protein or with Purkinje cell lysates caused animals to produce Yo-Ab, but did not result in disease [31, 32]. Adoptive transfer of mononuclear cells from the blood and CSF of Yo-Ab-positive patients into SCID mice or of Yo-antigen-sensitised mouse mononuclear cells has also failed to produce disease [33]. Recent studies have demonstrated the presence of cdr2-specific cytotoxic T lymphocytes in the blood and CSF of PCA patients [34, 35], showing that killer T cells are present in these patients and that these recognise the same antigen as Yo-Ab. Taken together, these results suggest that PCA with Yo-Ab is probably an autoimmune disease directed against cdr2 and may be due to T-cell-mediated destruction of the Purkinje cells, rather than to an antibody-mediated mechanism. However, further studies are still required in order to characterise the mechanisms leading to Purkinje cell death.

PCA with Anti-Hu Antibodies

Anti-Hu antibodies (Hu-Ab), the antibodies most frequently associated with PND [36], recognise a family of 35- to 40-kDa proteins found in the cytoplasm and nucleus of all neurons [37, 38]. Patients with Hu-Ab differ from those with Yo-Ab in terms of a frequent association with small-cell lung carcinoma (SCLC) [38], the absence of gynaecological tumour, and the presence of inflammatory CNS infiltrates and neuronal degeneration that is not restricted to cerebellar Purkinje cells [38]. The most frequent neurological symptoms associated with Hu-Ab are encephalomyelopathy and sensory neuronopathy [36, 38] (see Chap. 7). However, 13% of patients with Hu-Ab present with a subacute cerebellar syndrome that in the initial stages cannot be differentiated from PCA [38, 39].

Cerebellar ataxia occurs at the same frequency in male and female patients with Hu-Ab [39] and the average age of the patients is 66 years. The cerebellar ataxia precedes detection of the associated tumour in more than 80% of patients by a median of 3 months [39]. Cerebellar signs usually begin with symmetrical truncal and appendicular ataxia. Dysarthria and nystagmus are seen in more than 50% of cases, and diplopia is an early and frequent symptom [39]. Development of cerebellar ataxia is usually subacute and the patients require assistance in performing activities of daily living within less than 1 month. Extracerebellar symptoms are detectable in more than 80% of patients with cerebellar ataxia and Hu-Ab, often within the first month of the disease. The most frequent are encephalitis, brainstem or cranial nerve dysfunction, autonomic dysfunction, Lambert-Eaton myasthenic syndrome (LEMS) and pure sensory neuropathy, the last of which typically evolves asymmetrically and is seen in more than 40% of patients [39]. The CSF can be normal, but more typically a mild inflammation is observed. Cranial CT or MRI scans are initially normal; after 6 months, they can show signs of cerebellar atrophy in 36% of patients, but may remain normal even after 2 years of evolution [39]. The neurological symptoms are the cause of death in more than 65% of patients [39].

As in patients with Yo-Ab, there are several arguments in favour of immune system involvement in the neurological symptoms associated with Hu-Ab. The CSF of these patients contains lymphocytes and oligoclonal bands and there is intrathecal synthesis of Hu-Ab [18, 38, 39]. Deposits of Hu-Ab are seen in the patient's brain and tumour [40]. In addition, serum IgG from a patient with Hu-Ab has been shown to induce lysis of cultured cerebellar granule cells [41], although this result was not confirmed by another laboratory [42]. However, internalisation of Hu-Ab by neurons apparently does not occur in vivo [42, 43], and only low levels of bound complement are found in the nervous system of patients with Hu-Ab [44, 45]. Taken together, these results suggest that direct neurotoxicity induced by Hu-Ab does not play a crucial role in the neurological symptoms.

The Hu antigens have been cloned and found to belong to a family of neuronal RNA binding proteins that are highly homologous to a *Drosophila* protein, ELAV, which is crucial for the development of the nervous system in the fly [46-48]. The exact function of the Hu proteins is unknown, but their homology to ELAV and their early expression during the embryogenesis of the mammalian nervous system [49, 50] suggest that they are also crucial for the development and maintenance of the neuronal phenotype. The potential role of Hu proteins in PND was explored by immunising animals with recombinant Hu proteins [43, 51] but, although the animals developed high titres of Hu-Ab, no disease was seen. However, none of these studies modelled the intrathecal synthesis of Hu-Ab that occurs in patients. Interestingly, in one model involving Hu DNA immunisation of mice previously implanted with a neuroblastoma cell line constitutively expressing Hu proteins, the immunised mice showed significant tumour growth inhibition, and the tumours in immunised animals contained three times more inflammatory cells with a higher T lymphocyte ratio than those in control mice. This suggests direct involvement of T lymphocytes in the tumour growth inhibition [51]. In patients with Hu-Ab, as in the Yo syndrome, a few studies have suggested that cell-mediated autoimmunity may be the main mechanism involved in brain injury and anti-tumoral immunity. In the patient's brain, CD4+ and cytotoxic CD8+ T cells predominate in the interstitial infiltrates and are occasionally seen surrounding neurons [45]. Analysis of T cell receptors in the inflammatory infiltrates of the nervous system and tumour of patients with Hu-Ab showed limited T cell receptor usage and *in situ* oligoclonal expansion of CD8+ cytotoxic T cells [52], demonstrating that these lymphocytes are attracted by a specific antigen. Hu protein is a specific antigenic target for autoreactive memory-primed CD4+ T cells of patients with Hu-Ab, and these lymphocytes are presumably of the Th1 subtype [53]. Cytotoxic T cells directed against Hu proteins are also probably present in patients with Hu-Ab [54]. Taken together, these results suggest that the neurological symptoms seen in patients with Hu-Ab are the result of an autoimmune disease directed against Hu protein.

PCA with Anti-CV2 Antibodies

Anti-CV2 antibodies (CV2-Ab) were originally identified in a patient with PCA, uveitis and peripheral neuropathy [55]. This patient's serum did not contain any of the previously described antibodies associated with PND, but did contain antibodies directed against a 66-kDa developmentally regulated brain protein expressed in the adult brain by a subset of oligodendrocytes (Fig. 2) [56, 57]. Since this first case, 30 more have been identified and the incidence of CV2-Ab in PCA is estimated at about 7% [58].

Most patients with CV2-Ab are male (70%), with an average age of 62 years [56, 58]. The neurological disorders precede or are contemporary with the detection of cancer in most cases (80%). The most frequently associated tumour is SCLC (60%) [56, 58]; the other associated cancers are malignant thymoma and uterine sarcoma. The neurological disorders seen in patients with CV2-Ab differ from patient to patient. Many of the symptoms described in PND, such as limbic encephalitis, subacute sensory neuronopathy, cerebellar ataxia, LEMS and encephalomyelitis, are observed, with cerebellar ataxia being seen in 60% of cases.

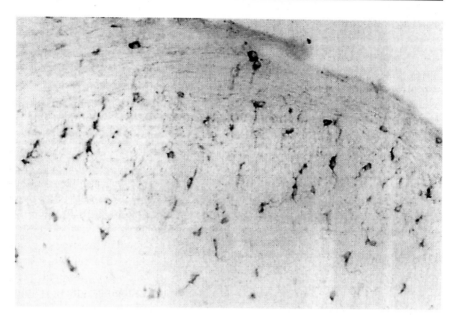

Fig. 2. Typical oligodendroglial immunoperoxidase staining of anti-CV2 antibodies in ventral spino-cerebellar and spinal trigeminal nerve tract of a 6-week-old rat brainstem. (Original magnification x 200)

As in other PND, onset is typically fairly abrupt and patients display signs and symptoms reflecting severe pancerebellar dysfunction, including nystagmus, dysarthria and severe appendicular gait ataxia. Retinopathy and optic neuritis are especially frequent in patients with CV2-Ab (present in 20%), and can be associated with cerebellar ataxia [55, 59]. Brain MR scans are usually normal. Inflammatory signs are present in the CSF in the majority of cases (70%) [56, 58].

The role of CV2-Ab in the pathology and the mechanisms involved in the cerebellar ataxia are unknown. However, as in other PCA types, an immune-mediated mechanism is suspected because of the inflammatory infiltrates seen in the cerebellum [39, 55]. The CV2 antigen has been cloned [60] and is a known to be member of a protein family recently described by other authors as ULIP (Unc-33-like proteins) or CRMP (collapsin response mediator proteins). This protein family is related to Unc-33, a nematode protein involved in axonal guidance [61], and might be involved in axonal outgrowth [62]; its role in PCA is not yet known.

In some cases, CV2-Ab can be associated in the same patient with Hu-Ab [58, 63]. Although less frequent than Hu-Ab, CV2-Ab seem to occur in patients with similar clinical pictures and tumours to those with Hu-Ab, suggesting a similar pathogenesis. Some clinical differences can be explained by differential expression of Hu and CV2 antigen in the nervous system [64].

Paraneoplastic Cerebellar Ataxia with Opsoclonus

Cerebellar ataxia is commonly seen in child and adult patients presenting a paraneoplastic opsoclonus-myoclonus (POM) syndrome [65, 66]. POM occurs in approximately 3% of children with neuroblastoma; in adults, it is less common and is mainly associated with lung cancer or breast carcinoma [66]. It typically has an abrupt onset with nausea, vomiting, ataxia of the trunk and limbs, opsoclonus and myoclonus. In some patients, opsoclonus occurs in the setting of pure, or relatively pure, pancerebellar dysfunction. There are no distinctive neuropathological abnormalities. Diffuse dropout of Purkinje cells, ranging from mild to almost complete, can be present, with or without perivascular mononuclear cell infiltrates [66] but, in a significant proportion of autopsied cases there are no identifiable histopathological abnormalities in the cerebellum or brainstem [67].

The serological abnormalities in adults with POM are heterogeneous. Some patients with opsoclonus, breast cancer and cerebellar ataxia have Yo-Ab and constitute a subset of those with the Yo syndrome [13]. Hu-Ab have been described in a few children with neuroblastoma and in a few adult patients with SCLC who developed opsoclonus and cerebellar ataxia as part of a multifocal encephalomyelitis [67]. Many patients either have no identifiable antibodies or have atypical anti-neuronal antibodies distinct from Hu-Ab or other recognised types [68, 69]. A separate group of patients with POM, breast cancer and ataxia (mainly truncal) has anti-Ri antibodies (Ri-Ab) [70]. In immunohistochemical tests, Ri-Ab stain nuclei of all neurons in the central nervous system in a pattern identical to that seen with Hu-Ab. However, Ri-Ab are distinguished by their binding to two groups of 50- and 80-kDa proteins on human neuronal immunoblots [70]. Autopsy of a single patient with opsoclonus, ataxia and Ri-Ab showed severe Purkinje cell loss and perivascular and interstitial infiltrates of B lymphocytes and CD4+ T lymphocytes in the brainstem. Direct immunohistochemistry showed intraneuronal IgG deposits, predominantly in the basal pons and dorsal midbrain [71].

Two Ri antigens, termed NOVA-1 and NOVA-2, have been cloned by screening a human cerebellar cDNA library with Ri-Ab [72, 73]. The NOVA proteins share sequence homology with a group of nuclear RNA-binding proteins distinct from Hu proteins and may be involved in the regulation of mRNA splicing, especially that of the glycine and GABA-A receptor pre-mRNAs [74, 75]. Some authors have demonstrated in vitro that affinity-purified Ri-Ab can block NOVA-1 binding to glycine receptor pre-RNA [76, 77] and may lead to aberrant regulation of glycine receptor expression. Interestingly, naturally occurring mutations of members of the glycine receptor family in both humans and mice lead to myoclonic neurological symptoms similar to those seen in POM [78, 79]. Thus, in some patients with opsoclonus and cerebellar ataxia, Ri-Ab may inhibit NOVA-1 RNA interactions and be responsible for the neurological symptoms.

PCA with Hodgkin's Disease

Of the cancers associated with PCA, Hodgkin's disease (HD) is the third most common after SCLC and ovarian cancer. The first pathologically confirmed case of PCA in association with HD was that reported by Malamud in 1957 [80]. Since that time, more than 50 other cases have been reported [17, 81-83]. HD-associated PCA occurs mainly in young men, with a median age of 30 years. In HD-associated PCA, the lymphoma precedes the ataxia by several months or years in 80% of patients, and ataxia often occurs during a prolonged complete remission [81]. In most patients, the abrupt or subacute onset of gait ataxia is the most common initial complaint. Others experience diplopia, oscillopsia, vertigo, or a "flu-like" syndrome with headache as their initial symptom. Neurological findings are symmetrical and confined to the cerebellum in most patients, but occasionally encephalopathy, Babinski reflexes or mild sensory neuropathy can be seen. Downbeat nystagmus is often observed, as in Yo-syndrome.

Anti-Purkinje cell antibodies have been identified in 30% of patients with HD-associated PCA [81]. The staining pattern of these anti-Purkinje cell antibodies is distinct from that of Yo-Ab or Hu-Ab, and different subtypes have been described. One of these, called Tr-Ab, has been particularly studied [83]. In the human cerebellum, Tr-Ab label the cytoplasm and proximal dendrites of Purkinje cells with a pattern of labelling similar to that seen with Yo-Ab. However, in contrast to the results with Yo-Ab, the molecular layer is diffusely stained, showing multiple positive small dots, suggestive of immunoreactivity with the dendritic spines of the Purkinje cells (Fig. 3). Immunoblotting failed to detect a band reactive with

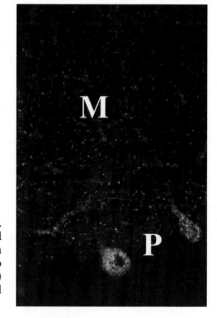

Fig. 3. Typical staining of anti-Tr antibodies. Anti-Tr antibodies labeled the cytoplasm and proximal dendrites of Purkinje cells (*P*) with a pattern similar to that observed with anti-Yo antibodies. However, the molecular layer (*M*) shows a typical fine dotted pattern. (Original magnification x 400)

Tr-Ab; however, the abolition of Tr-Ab immunoreactivity by pre-incubation of the tissue sections with pepsin suggests that the antigen is probably a protein and that Tr-Ab may be directed against a conformational epitope [83, 84]. Tr-Ab were not found in a large series of patients with cerebellar disorders or isolated HD [83], suggesting that they are probably highly specific for HD-associated PCA and may play a role in the cerebellar ataxia.

Recently, antibodies reacting specifically with the metabotropic glutamate receptor mGluR1 (mGluR1-Ab) have been described in two patients with HD-associated PCA [82]. Interestingly, mGluR1-Ab have not been described in other groups of patients, and the purified IgG from the serum of both patients blocked the glutamate-stimulated formation of inositol phosphates in Chinese hamster ovary cells expressing mGluR1, and injection of IgG from the serum or CSF into the cerebellar subarachnoid space of mice caused severe reversible ataxia [82]. These results indicate, for the first time, that anti-neuronal antibodies can cause CNS disease by blocking neuronal receptors. However, this mechanism cannot explain HD-associated PCA in the absence of mGluR1-Ab.

PCA with Anti-Voltage-Gated Calcium Channel Antibodies

Anti-voltage-gated calcium channel antibodies (VGCC-Ab), especially those directed against the P/Q type VGCC, are mainly associated with paraneoplastic and non-paraneoplastic LEMS [85, 86]. However, some studies have demonstrated that the frequency of cerebellar ataxia in patients with LEMS is higher than that expected by chance, and that LEMS with ataxia is usually associated with cancer [87]. VGCC-Ab against the P/Q type VGCC have also been found in patients with SCLC and PCA without LEMS [39]. P/Q type VGCCs have been described in the central nervous system, especially in the cerebellum [88]. Furthermore, gene mutations coding for VGCCs have been implicated in certain familial cerebellar ataxias [88], and in vitro VGCC-Ab can reduce P- and Q-type calcium currents in cerebellar neurons [90]. Taken together, these data suggest that VGCC-Ab could play a direct role in certain PCA.

PCA with Atypical Antibodies or Without Antibodies

Some cases of PCA have been reported in association with other anti-neuronal antibodies, such as anti-amphiphysin (Fig. 4) [91], anti-Ma1 [92] or atypical anti-Purkinje cell antibodies [93, 94]. The significance of most of these antibodies is uncertain, but they indicate the existence of immune mechanisms which may be involved in the neurological dysfunction. However, a high percentage (probably more than 50%) of patients with confirmed PCA do not have demonstrable circulating anti-neuronal antibodies [17, 39]. The aetiological factors involved in PCA in patients without antibodies remain speculative, but may also be autoimmune, as many of these patients have CSF inflammation [39]. However, the clinical evolution of sero-positive and sero-negative patients seems to be different [39], suggesting that there may be multiple mechanisms resulting in PCA.

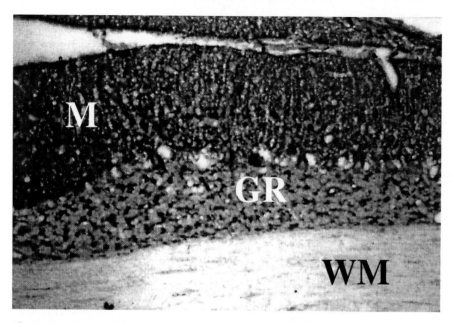

Fig. 4. Typical immunoperoxidase staining of anti-amphiphysin antibodies in adult rat cerebellum. A diffuse labeling of the neuropil is observed in the molecular layer (*M*) with a fine granular staining around granule (*GR*) and Purkinje cells. The white matter (*WM*) is always negative. (Original magnification x 200)

Non-Paraneoplastic Cerebellar Ataxia

Recently, autoimmunity has also been suspected to be involved in certain cases of cerebellar ataxia of non-paraneoplastic origin (Table 2). Two groups of patients have now been clearly defined, but other groups may be identified in the future.

Table 2. Main subtypes of immune-mediated N-PCA

Immune-mediated cerebellar ataxia	Associated antibodies	Clinical presentation
GAD-Ataxia	Anti-GAD	Mainly female patients Average age 55 years Progressive cerebellar ataxia with nystagmus Association with polyendocrinopathy CSF oligoclonal IgG bands
Gluten-Ataxia	Anti-Gliadin	Mainly male patients Average age 54 years Progressive cerebellar ataxia with axonal neuropathy Gastrointestinal symptoms HLA DQ2

N-PCA with Anti-Glutamic Acid Decarboxylase Antibodies

GAD, a major enzyme of the nervous system, catalyses the conversion of glutamate to GABA. This enzyme, also expressed by pancreatic β-cells, has been identified as a dominant and essential autoantigen in the development of insulin-dependent diabetes mellitus (IDDM) [95]. GAD-Ab are present in up to 80% of newly diagnosed IDDM patients and can be detected many years before clinical onset of the disease [96]. High levels of GAD-Ab are found in the serum and CSF of at least 60% of patients with the stiff-person syndrome, a rare disorder of the central nervous system characterised by progressive muscle rigidity with superimposed painful spasms [97, 98]. Stiff-person syndrome patients with GAD-Ab usually present with IDDM and other organ-specific autoimmune manifestations, suggesting that the stiff-man syndrome may have an autoimmune-mediated pathogenesis [99] (see Chapter 9). Recent studies have shown that high levels of GAD-Ab may also be found in a few patients with cerebellar ataxia (Fig. 5), suggesting that in these cases the cerebellar syndrome might also have an autoimmune pathogenesis [100-105].

Most patients with GAD-Ab and cerebellar ataxia are female (90%), median age is 55 years at the onset of the cerebellar syndrome. A family history of autoimmune diseases, such as IDDM or thyroiditis, is common [104]. Late-onset IDDM is seen in 71% of patients, and other autoimmune disorders, such as thyroiditis (Hashimoto's or Grave's disease), pernicious anaemia, myasthenia gravis, thymoma, psoriasis and coeliac disease, have also been reported [102-105]. The cerebellar symptoms progress slowly, suggesting a degenerative disease. A subacute onset has been seen in only one case [102]. The main cerebellar sign is gait ataxia. Limb ataxia is also observed later in the disease course. After several years of disease,

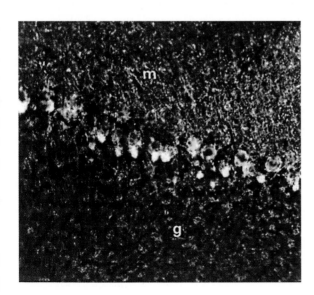

Fig. 5. Typical staining of anti-GAD antibodies. A prominent accumulation of immunoreactivity is observed in the nerve terminals of Basket cells around the initial axons segment of Purkinje cells with a fine dotted staining in molecular (*m*) and granular layer (*g*). (Original magnification x 200)

the cerebellar symptoms can prevent a completely independent way of life in 70% of patients. Nystagmus and dysarthria are frequently observed. Other neurological symptoms can be seen in some patients with cerebellar ataxia and GAD-Ab; these include leg rigidity suggesting a localised stiff-person syndrome [101, 104, 106], peripheral neuropathy [100] and myasthenia gravis [102]. The CSF examination is generally normal in terms of cell numbers and protein level; however, oligoclonal IgG bands, not seen in the serum, are frequently detected by isoelectric focusing and immunoblotting of IgG [104]. Brain MRI shows pure cerebellar atrophy in 50% of patients, but brainstem atrophy has never been observed [104].

The presence of oligoclonal bands in the CSF, GAD-Ab intrathecal synthesis and organ-specific autoimmune disorders in most patients with cerebellar ataxia and GAD-Ab, clearly distinguishes them from patients with other non-familial late-onset cerebellar ataxias and suggests an autoimmune basis. Furthermore, the co-occurrence of stiff-person syndrome and cerebellar ataxia with high titres of GAD-Ab [101, 104, 106] suggests an overlap between the two neurological symptoms and common pathological mechanisms. Recent studies have suggested that GAD-Ab from patients with stiff-person syndrome may be pathogenic, because, unlike the GAD-Ab from patients with IDDM, they can reduce GAD enzyme activity and GABA synthesis [107]. In one case of cerebellar ataxia, GAD-Ab was shown to produce selective suppression of GABAergic transmission in in vitro experiments using isolated rat cerebellar slices [105].

N-PCA with Gluten Ataxia

"Gluten ataxia" refers to a condition recently described in a group of patients with idiopathic late-onset cerebellar ataxia and gluten sensitivity, defined by the presence of circulating antibodies (IgG or IgA) to gliadin [108].

In contrast to GAD-Ab positive patients, patients with gluten ataxia are predominantly male (65%), with a median age of 54 years at the onset of the cerebellar syndrome [108, 109]. There is no family history of spinocerebellar degeneration. Gastrointestinal symptoms are not major clinical signs, being seen in only 40% of patients after specific questioning (mainly abdominal distension, abdominal pain or diarrhoea). A distal duodenal biopsy is normal in more than 50% of patients, but can show lymphocytic infiltration or the typical villous atrophy of coeliac disease [108]. The HLA DQ2 genotype is found in the majority of patients (82%). Typically, the cerebellar symptoms progress slowly, suggesting a degenerative disease, and become severe after several years. The main cerebellar sign is gait ataxia. Limb ataxia is seen in some cases; if present, it is more marked in the lower limbs. Pyramidal, extrapyramidal or autonomic features are not observed. In contrast to patients with GAD-Ab and cerebellar ataxia, nystagmus is occasionally seen, and axonal sensory-motor peripheral neuropathy, expressed clinically by absent deep tendon reflexes and distal sensory loss, is frequent (70%) [108]. CT and brain MRI show cerebellar atrophy affecting both the vermis and the hemispheres in 20% of patients, but brainstem atrophy has never been observed [108].

The physiopathology of gluten ataxia is unknown. However, T cell infiltration of the cerebellum, in which patchy loss of Purkinje cells is seen, and of the peripheral nervous system [108], suggests immunologically mediated neural damage.

Other Potentially Immune-Mediated N-PCA

An autoimmune mechanism of cerebellar ataxia has been suggested in other sporadic cases of cerebellar degeneration associated with anti-GluR2 [110], anti-Purkinje cell [111], anti-VGCC (see Chapter 1) or anti-MPP1 [112] antibodies. However, the number of described observations is as yet too small for definitive conclusions about the ataxia mechanism in these cases.

Conclusions

These data allow several conclusions to be drawn. Firstly, despite the blood-brain-barrier, an immune reaction may play an important role in certain cerebellar ataxias. Secondly, although no studies have yet conclusively proved that the associated antibodies are pathogenic, they can be employed as useful diagnosis markers to classify subtypes of cerebellar ataxia. PCA is the best studied example. However, autoimmune-mediated cerebellar ataxia can also be seen in patients without cancer. Thirdly, a cytotoxic T lymphocyte response could play an important role not only during the acute stages of cerebellar ataxia, but also during the progressive cerebellar degeneration, and should be carefully studied. Finally, although autoimmune-mediated cerebellar ataxias are rare and clearly not a major health problem, the elucidation of the pathogenesis of these syndromes may have important implications in our understanding of neuronal degeneration and brain immunology.

References

1. Posner JB, Furneaux HM (1990) Paraneoplastic syndromes. In: Waksman BH (ed) Immunologic mechanisms in neurologic and psychiatric diseases. Raven Press, New York, pp 187-219
2. Brouwer B (1919) Beitrag zur Kenntnis der chronischen diffusen Kleinhirnerkrankungen. Neurol Centralbl 38:674-682
3. Brain WR, Daniel PM, Greenfield JG (1951) Subacute cortical cerebellar degeneration and its relation to carcinoma. J Neurol Neurosurg Psychiatry 14:59-75
4. Henson RA, Urich H (1982) Paraneoplastic disorders. In: Henson RA, Urich H, (ed) Cancer and the nervous system. Blackwell Scientific Publication, Oxford, pp 311-621
5. Anderson NE, Cunningham JM, Posner JB (1987) Autoimmune pathogenesis of paraneoplastic neurological syndromes. CRC Crit Rev Neurobiol 3:245-299
6. Verschuuren J, Chuang L, Rosenblum MK et al (1996) Inflammatory infiltrates and complete absence of Purkinje cells in anti-Yo associated paraneoplastic cerebellar degeneration. Acta Neuropathol 91:519-525
7. Russell DS (1961) Encephalomyelitis and carcinomatous neuropathy. In: Van Bogaert

L, Radermecker J, Hozay J, Lowenthal A (eds) The encephalitides. Elsevier, Amsterdam, pp 131-135

8. Trotter JL, Hendin BA, Osterland CK (1976) Cerebellar degeneration with Hodgkin disease. Arch Neurol 33:660-661

9. Dalmau JO, Posner JB (1999) Paraneoplastic syndromes. Arch Neurol 56:405-408

10. Keime-Guibert F, Graus F, Fleury A et al (2000) Treatment of paraneoplastic neurological syndromes with anti-neuronal antibodies (anti-Hu, anti-Yo) with a combination of immunoglobulins, cyclophosphamide and methylprednisolone. J Neurol Neurosurg Psychiatry 68:479-482

11. Greenlee JE, Brashear HR (1983) Antibodies to cerebellar Purkinje cells in patients with paraneoplastic cerebellar degeneration and ovarian carcinoma. Ann Neurol 14:609-613

12. Jaeckle KA, Graus F, Houghton A et al (1985) Autoimmune response of patients with paraneoplastic degeneration to a Purkinje cell cytoplasmic protein antigen. Ann Neurol 18:592-600

13. Peterson K, Rosenblum MK, Kotanides H, Posner JB (1992) Paraneoplastic cerebellar degeneration. I. A clinical analysis of 55 anti-Yo antibody-positive patients. Neurology 42:1931-1937

14. Rojas I, Graus F, Keime-Guibert F et al (2000) Long-term clinical outcome of paraneoplastic cerebellar degeneration and anti-Yo antibodies. Neurology 55:713-715

15. Felician O, Renard JL, Vega F et al (1995) Paraneoplastic cerebellar degeneration with anti-Yo antibody in man. Neurology 45:1226-1227

16. Krakauer J, Balmaceda C, Gluck JT et al (1996) Anti-Yo-associated paraneoplastic cerebellar degeneration in a man with adenocarcinoma of unknown origin. Neurology 46:1486-1487

17. Anderson NE, Rosenblum MK, Posner JB (1988) Paraneoplastic cerebellar degeneration: clinical-immunological correlation. Ann Neurol 24:559-567

18. Furneaux HM, Reich L, Posner JB (1990) Autoantibody synthesis in the central nervous system of patients with paraneoplastic syndromes. Neurology 40:1085-1091

19. Giometto B, Marchiori GC, Nicolao P et al (1997) Subacute cerebellar degeneration with anti-Yo autoantibodies: immunohistochemical analysis of the immune reaction in the central nervous system. Neuropathol Appl Neurobiol 23:468-474

20. Chen YT, Rettig WJ, Yenamandra AK et al (1990) Cerebellar degeneration related (CDR) antigen: a highly conserved neuroectodermal marker mapped to chromosomes X in human mouse. Proc Natl Acad Sci 87:3077-3081

21. Fatallah-Shaykh H, Wolf S, Wong E et al (1991) Cloning of a leucine-zipper protein recognized by the sera of patients with antibody-associated paraneoplastic cerebellar degeneration. Proc Natl Acad Sci USA 88:3451-3454

22. Sakai K, Mitchell DJ, Tsukamoto T, Steinman L (1990) Isolation of a complementary DNA clone encoding an autoantigen recognized by an anti-neuronal cell antibody from a patient with paraneoplastic cerebellar degeneration. Ann Neurol 28:692-698

23. Darnell JC, Albert ML, Darnell RB (2000) Cdr2, a target antigen of naturally occurring human tumor immunity, is widely expressed in gynecological tumors. Cancer Res 60:2136-2139

24. Okano HJ, Park WY, Corradi JP, Darnell RB (1999) The cytoplasmic Purkinje onconeural antigen cdr2 down-regulates c-Myc function: implications for neuronal and tumor cell survival. Genes Dev 16:2087-2097

25. Sakai K, Ogasawara T, Hirose G et al (1993) Analysis of autoantibody binding to 52-kd paraneoplastic cerebellar degeneration-associated antigen expressed in recombinant proteins. Ann Neurol 33:373-380

26. Borges LF, Elliot PJ, Gill R et al (1985) Selective extraction of small and large molecules from the cerebrospinal fluid by Purkinje neurons. Science 228:346-348
27. Fabian RH, Petroff G (1987) Intraneuronal IgG in the central nervous system: uptake by retrograde axonal transport. Neurology 37:1780-1784
28. Graus F, Illa I, Agusti M, Ribalta T et al (1991) Effect of intraventricular injection of anti-Purkinje cell antibody (anti-Yo) in a guinea pig model. J Neurol Sci 106:82-87
29. Jaeckle KA, Stroop WG (1986) Intraventricular injection of paraneoplastic anti-Purkinje cell antibody in a rat model. Neurology 36 (Suppl 1):332
30. Greenlee JE, Burns JB, Rose JW et al (1995) Uptake of systematically administered human anti-cerebellar antibody by rat Purkinje cells following blood brain barrier disruption. Acta Neuropathol 89:341-345
31. Sakai K, Gofuku M, Kitagawa Y et al (1995) Induction of anti-Purkinje cell antibodies in vivo by immunizing with a recombinant 52 kDa paraneoplastic cerebellar degeneration associated protein. J Neuroimmunol 60:135-141
32. Tanaka M, Tanaka K, Onoreda O, Tsuji S (1995) Trial to establish an animal model of paraneoplastic cerebellar degeneration with anti-Yo antibody: 1. Mouse strains bearing different MHC molecules produce antibodies on immunization with recombinant Yo protein, but do not cause Purkinje cell loss. Clin Neurol Neurosurg 97:95-100
33. Tanaka K, Tanaka M, Igarashi S et al (1995) Trial to establish an animal model of paraneoplastic cerebellar degeneration with anti-Yo antibody: 2. Passive transfer of murine mononuclear cells activated with recombinant Yo protein to paraneoplastic cerebellar degeneration lymphocytes in severe combined immunodeficiency mice. Clin Neurol Neurosurg 97:101-105
34. Albert ML, Darnell JC, Bender A et al (1998) Tumor-specific killer cells in paraneoplastic cerebellar degeneration. Nature Med 11:1321-1324
35. Albert ML, Austin L, Darnell RB (2000) Detection and treatment of activated T cells in the cerebrospinal fluid of patients with paraneoplastic cerebellar degeneration. Ann Neurol 47:9-17
36. Voltz RD, Posner JB, Dalmau J, Graus F (1997) Paraneoplastic encephalomyelitis: an update of the effects of the anti-Hu immune response on the nervous system and tumour. J Neurol Neurosurg Psychiatry 63:133-136
37. Graus F, Cordon-Cardo C, Posner JB (1985) Neuronal antinuclear antibody in sensory neuronopathy from lung cancer. Neurology 35:538-543
38. Dalmau J, Graus F, Rosenblum MK, Posner JB (1992) Anti-Hu associated paraneoplastic encephalomyelitis/sensory neuropathy. Medicine 71:59-72
39. Mason WP, Graus F, Lang B et al (1997). Paraneoplastic cerebellar degeneration and small cell lung cancer. Brain 120:1279-1300
40. Dalmau J, Furneaux HM, Rosenblum MK et al (1991) Detection of the anti-Hu antibody in specific regions of the nervous system and tumor from patients with paraneoplastic encephalomyelitis and sensory neuropathy. Neurology 41:1757-1764
41. Greenlee JE, Parks TN, Jaeckle KA (1993) Type IIa (anti-Hu) antineuronal antibodies produce destruction of rat cerebellar granule neurons in vitro. Neurology 43:2049-2054
42. Hormigo A, Lieberman F (1994) Nuclear localization of anti-Hu antibody is not associated with in vitro cytotoxicity. J Neuroimmunol 55:205-212
43. Sillevis Smitt PAE, Manley GT, Posner JB (1995) Immunization with the paraneoplastic encephalomyelitis antigen HuD does not cause neurologic disease in mice. Neurology 45:1873-1878
44. Panegyres PK, Reading MC, Esiri MM (1993) The inflammatory reaction of paraneoplastic ganglionitis and encephalitis: an immunohistochemical study. J Neurol 240:93-97

45. Jean WC, Dalmau J, Ho A, Posner JB (1994) Analysis of the IgG subclass distribution and inflammatory infiltrates in patients with anti-Hu associated paraneoplastic encephalomyelitis. Neurology 44:140-147

46. Szabo A, Dalmau J, Mauley G et al (1991) HuD, a paraneoplastic encephalomyelitis antigen, contains RNA-binding domains and is homologous to ELAV and sex-lethal. Cell 67:325-333

47. Dropcho EJ, King PH (1994) Autoantibodies against the Hel-N1 RNA-binding protein among patients with lung carcinoma: an association with type I anti-neuronal nuclear antibodies. Ann Neurol 36:200-205

48. Sakai K, Gofuku M, Kitagawa Y et al (1994) A hippocampal protein associated with paraneoplastic neurologic syndrome and small cell lung carcinoma. Biochem Biophys Res Commun 199:1200-1208

49. Marusich M, Furneaux H, Henion P, Weston J (1994) Hu neuronal proteins are expressed in proliferating neurogenic cells. J Neurobiol 25:143-155

50. Akamatsu W, Okano HJ, Osumi N et al (1999) Mammalian ELAV-like neuronal RNA-binding proteins HuB and HuC promote neuronal development in both the central and the peripheral nervous systems. Proc Natl Acad Sci 96:9885-9890

51. Carpentier AF, Rosenfeld MR, Delattre JY et al (1998) DNA vaccination with HuD inhibits growth of a neuroblastoma in mice. Clin Cancer Res 4:2819-2824

52. Voltz RD, Dalmau J, Posner JB, Rosenfeld MR (1998) T-cell receptor analysis in anti-Hu associated paraneoplastic encephalomyelitis. Neurology 51:1146-1150

53. Benyahia B, Liblau R, Merle-Béral H et al (1999) Cell-mediated autoimmunity in para-neoplastic neurological syndromes with anti-Hu antibodies. Ann Neurol 45:162-167

54. Tanaka K, Tanaka M, Inuzuka T et al (1999) Cytotoxic T lymphocyte-mediated cell death in paraneoplastic sensory neuropathy with anti-Hu antibody. J Neurol Sci 163:159-162

55. Antoine JC, Honnorat J, Koenig F et al (1993) Posterior uveitis and paraneoplastic encephalomyelitis with auto-antibodies reacting against cytoplasmic proteins of brain and retina. J Neurol Sci 117:215-223

56. Honnorat J, Antoine JC, Derrington E et al (1996) Antibodies to a subpopulation of glial cells and a 66 kD developmental protein in patients with paraneoplastic neuro-logical syndromes. J Neurol Neurosurg Psychiatry 61:270-278

57. Honnorat J, Aguera M, Zalc B et al (1998) POP66, a paraneoplastic encephalomyelitis related antigen is a marker of adult oligodendrocytes. J Neuropathol Exp Neurol 57:311-322

58. Rogemond V, Honnorat J (2000) Anti-CV2 autoantibodies and paraneoplastic neuro-logical syndromes. Clin Rev Allergol Immunol 19:48-57

59. De la Sayette V, Bertran F, Honnorat J et al (1998) Paraneoplastic cerebellar syndrome and optic neuritis with anti-CV2 antibodies: clinical response to excision of the pri-mary tumor. Arch Neurol 55:405-408

60. Honnorat J, Byk T, Kusters I et al (1999) Ulip/CRMP proteins are recognized by autoan-tibodies in paraneoplastic neurological syndromes. Eur J Neurosci 11:4226-4232

61. Wang LH, Strittmatter SM (1996) A family of rat CRMP genes is differentially expressed in the nervous system. J Neurosci 16:6197-207

62. Goshima Y, Nakamura F, Strittmatter P, Strittmatter SM (1995) Collapsin-induced growth cone collapse mediated by an intracellular protein related to UNC-33. Nature 376:509-14

63. Honnorat J, Guillon B, De Ferron E et al (1997) Association of anti-neural autoanti-bodies in a patient with paraneoplastic cerebellar syndrome and small cell lung carci-noma. J Neurol Neurosurg Psy 62:425-426

64. Antoine JC, Honnorat J, Camdessanche JP et al (2001) Paraneoplastic anti-CV2 antibodies react with peripheral nerve and are associated with a peripheral neuropathy different from that of anti-Hu syndromes. Ann Neurol 49:214-221

65. Telander RL, Smithson WA, Groover RV (1989) Clinical outcome in children with acute cerebellar encephalopathy and neuroblastoma. J Pediatr Surg 24:11-14

66. Anderson NE, Budde-Steffen C, Rosenblum MC et al (1988) Opsoclonus, myoclonus, ataxia and encephalopathy in adults with cancer: a distinct paraneoplastic syndrome. Medicine (Baltimore) 67:100-109

67. Hersh B, Dalmau J, Dangond F et al (1994) Paraneoplastic opsoclonus-myoclonus associated with anti-Hu antibody. Neurology 44:1754-1755

68. Cher LM, Hochberg FH, Teruya J et al (1995) Therapy for paraneoplastic neurological syndromes in six patients with protein A column immunoadsorption. Cancer 75:1678-1683

69. Honnorat J, Trillet M, Antoine JC et al (1997) Paraneoplastic opsomyoclonus, cerebellar ataxia and encephalopathy associated with anti-Purkinje cell antibodies. J Neurol 244:333-339

70. Luque FA, Furneaux HM, Ferziger R et al (1991) Anti-Ri: an autoantibody associated with paraneoplastic opsoclonus and breast cancer. Ann Neurol 29:241-251

71. Hormigo A, Dalmau J, Rosenblum MK et al (1994) Immunological and pathological study of anti-Ri associated encephalopathy. Ann Neurol 36:896-902

72. Buckanovich ARJ, Yang YYL, Darnell RB (1993) The onconeural antigen Nova-1 is a neuron RNA-binding protein and is specifically expressed in the developing motor system. Neuron 11:657-672

73. Yang YL, Lin Yin G, Darnell RB (1998) The neuronal RNA-binding protein Nova-2 is implicated as the autoantigen targeted in POMA patients with dementia. Proc Natl Acad Sci USA 95:13254-13259

74. Jensen KB, Dredge BK, Stefani G et al (2000) Nova-1 regulates neuron-specific alternative splicing and is essential for neuronal viability. Neuron 25:359-371

75. Polyrides AD, Okano HJ, Yang YY et al (2000) A brain-enriched polyrimidine tract-binding protein antagonizes the ability of NOVA to regulate neuron-specific alternative splicing. Proc Natl Acad Sci USA 97:6350-6355

76. Buckanovich ARJ, Posner JB, Darnell RB (1996) Nova, the paraneoplastic Ri antigen is homogous to an RNA-binding protein, the activity of which is inhibited by paraneoplastic antibodies. J Neurosci 16:1114-1122

77. Buckanovich RJ, Darnell RB (1997) The neuronal RNA binding protein NOVA 1 recognizes specific RNA targets in vitro and in vivo. Mol Cell Biol 17:3194-3201

78. Ryan SG, Buckwalter MS, Lynch JW et al (1994) A missense mutation in the gene encoding the alpha 1 subunit of the inhibitory glycine receptor in the spasmodic mouse. Nat Genet 7:131-135

79. Saul B, Schmieden V, Kling C et al (1994) Point mutation of glycine receptor alpha 1 subunit in the spasmodic mouse affects agonist responses. FEBS Lett 350:71-76

80. Malamud N (1957) Atlas of neuropathology. University of California Press, Berkeley, pp 118

81. Hammack J, Kotanides H, Rosenblum MK, Posner JB (1992) Paraneoplastic cerebellar degeneration. II. Clinical and immunologic findings in 21 patients with Hodgkin's disease. Neurology 42:1938-1943

82. Sillevis-Smitt P, Kinoshita A, De Leeuw B et al (2000) Paraneoplastic cerebellar ataxia due to autoantibodies against a glutamate receptor. N Engl J Med 342:21-27

83. Graus F, Dalmau J, Valldeoriola F et al (1997) Immunological characterization of a

neuronal antibody (anti-Tr) associated with paraneoplastic cerebellar degeneration and Hodgkin's disease. J Neuroimmunol 74:55-61

84. Graus F, Gultekin SH, Ferrer I et al (1998) Localization of the neuronal antigen recognized by anti-Tr antibodies from patients with paraneoplastic cerebellar degeneration and Hodgkin's disease in the rat nervous system. Acta Neuropathol 96:1-7

85. Motomura M, Lang B, Johnston I et al (1997) Incidence of serum anti-P/O-type and anti-N-type calcium channel autoantibodies in the Lambert-Eaton myasthenic syndrome. J Neurol Sci 147:35-42

86. Lennon VA, Kryzer TJ, Griesmann GE et al (1995) Calcium-channel antibodies in the Lambert-Eaton syndrome and other paraneoplastic syndromes. N Engl J Med 332:1467-1474

87. Clouston PD, Saper CB, Arbizu T et al (1992) Paraneoplastic cerebellar degeneration. III. Cerebellar degeneration, cancer, and the Lambert-Eaton myasthenic syndrome. Neurology 42:1944-1950

88. Greenberg DA (1997) Calcium channels in neurological disease. Ann Neurol 42:275-282

89. Restituito S, Thompson RM, Eliet J et al (2000) The polyglutamine expansion in spinocerebellar ataxia type 6 causes a beta subunit-specific enhanced activation of P/Q-type calcium channels in Xenopus oocytes. J Neurosci 20:6394-6403

90. Pinto A, Gillard S, Moss F et al (1998) Human autoantibodies specific for the alpha 1A calcium channel subunit reduce both P-type and Q-type calcium currents in cerebellar neurons. Proc Natl Acad Sci 95:8328-8333

91. Antoine JC, Absi L, Honnorat J et al (1999) Anti-amphiphysin antibodies are associated with various paraneoplastic neurological syndromes and tumours. Arch Neurol 56:172-177

92. Dalmau JO, Gultekin SH, Voltz R et al (1999) Ma1, a novel neuron- and testis-specific protein, is recognized by the serum of patients with paraneoplastic neurological disorders. Brain 122:27-39

93. Darnell RB, Furneaux HM, Posner JB (1991) Antiserum from a patient with cerebellar degeneration identifies a novel protein in Purkinje cells, cortical neurons, and neuroectodermal tumors. J Neurosci 11:1224-1230

94. Tanaka K, Yamazaki M, Sata S et al (1986) Antibodies to brain proteins in paraneoplastic cerebellar degeneration. Neurology 36:1169-1172

95. Yoon JW, Yoon CS, Lim HW et al (1999) Control of autoimmune diabetes in NOD mice by GAD expression or suppression in beta cells. Science 284:1183-1187

96. Baekkeskov S, Aanstoot HJ, Christgau S et al (1990) Identification of the 64K autoantigen in insulin-dependent diabetes as the GABA-synthesizing enzyme glutamic acid decarboxylase. Nature 347:151-156

97. Solimena M, Folli F, Aparisi R et al (1990) Autoantibodies to GABA-ergic neurons and pancreatic beta cells in stiff-man syndrome. N Engl J Med 322:1555-1560

98. Grimaldi LM, Martino G, Braghi S et al (1993) Heterogeneity of autoantibodies in stiff-man syndrome. Ann Neurol 34:57-64

99. Solimena M, De Camilli P (1991) Autoimmunity to glutamic acid decarboxylase (GAD) in stiff-man syndrome and insulin-dependent diabetes mellitus. Trends Neurosci 14:452-457

100. Honnorat J, Trouillas P, Thivolet C et al (1995) Autoantibodies to glutamate decarboxylase in a patient with cerebellar cortical atrophy, peripheral neuropathy, and slow eye movements. Arch Neurol 52:462-468

101. Giometto B, Miotto D, Faresin F et al (1996) Anti-GABAergic neuron autoantibodies in a patient with stiff-man syndrome and ataxia. J Neurol Sci 143:57-59

102. Saiz A, Arpa J, Sagasta A et al (1997) Autoantibodies to glutamic acid decarboxylase in three patients with cerebellar ataxia, late-onset insulin dependent diabetes mellitus, and polyendocrine autoimmunity. Neurology 49:1026-1030
103. Abele M, Weller M, Mescheriakov S et al (1999) Cerebellar ataxia with glutamic acid decarboxylase autoantibodies. Neurology 52:857-859
104. Honnorat J, Saiz A, Giometto B et al (2001) Cerebellar ataxia with anti-glutamic acid decarboxylase antibodies: study of a series of 14 patients. Arch Neurol 58:225-230
105. Ishida K, Mitoma H, Song SY et al (1999) Selective suppression of cerebellar GABAergic transmission by an autoantibody to glutamic acid decarboxylase. Ann Neurol 46:263-267
106. Brashear HR, Login IS, Mathe SA, Phillips LH (1997) Cerebellar disorder in stiff-man syndrome [abstract]. Neurology 48:A433
107. Dinkel K, Meinck HM, Jury KM et al (1998) Inhibition of γ-aminobutyric acid synthesis by glutamic acid decarboxylase autoantibodies in stiff-man syndrome. Ann Neurol 44:194-201
108. Hadjivassiliou M, Grünewald RA, Chattopadhyay AK et al (1998) Clinical, radiological, neurophysiological, and neuropathological characteristics of gluten ataxia. Lancet 352:1582-1585
109. Pellecchia MT, Scala R, Filla A et al (1999) Idiopathic cerebellar ataxia associated with celiac disease: lack of distinctive neurological features. J Neurol Neurosurg Psychiatry 66:32-35
110. Gahring LC, Rogers SW, Twyman RE (1997) Autoantibodies to glutamate receptor subunit GluR2 in non familial olivopontocerebellar degeneration. Neurology 48:494-500
111. Ito H, Sayama S, Irie S et al (1994) Antineuronal antibodies in acute cerebellar ataxia following Epstein-Barr virus infection. Neurology 44:1506-1507
112. Fritzler MJ, Kerfoot SM, Feasby TE et al (2000) Autoantibodies from patients with idiopathic ataxia bind to M-phase phosphoprotein-1 (MPP1). J Invest Med 48:28-39

Chapter 9

Stiff-Man Syndrome: Pathogenetic, Nosological and Therapeutic Considerations

F. Folli[1], G. Piccolo[2]

Introduction

The acronym SMS (Stiff-Man Syndrome) identifies a syndrome of slowly progressive stiffness involving skeletal muscles (mainly axial) with superimposed muscle spasms. It was first described by Moersch and Woltmann in 1956 [1]. A set of diagnostic criteria was proposed by Gordon et al. [2] and Lorish et al. [3]. Diagnostic criteria included: 1) a prodrome of stiffness and rigidity in axial muscles; 2) a slow progression of stiffness involving proximal limb muscles, making walking difficult; 3) a fixed deformity, usually lordosis, of the spine; 4) the presence of superimposed muscle spasms, often precipitated by external stimuli; 5) normal motor and sensory nerve findings; 6) normal mental status and 7) an EMG finding of continuous motor unit activity (CMUA) at rest, abolished by intravenous diazepam or reduced by orally administered diazepam. It has been recently debated whether SMS is a single disease entity. Thirty-eight reported SMS patients were rexamined [4]; only seven of them (18.4%) fulfilled the Lorish diagnostic criteria [3].

Many had prevailing limb stiffness, associated signs of central or peripheral nervous system involvement, and some of them did not show superimposed muscle spasm, or did not have typical EMG findings. At that time, a "lumping" diagnostic tendency prevailed, so we suggested the possibility of splitting the SMS diagnosis into a "typical" form and "encephalomyelopathic" variants [4]. This has also been proposed by Brown and Marsden in their recent review of clinicopathological, immunological and electrophysiological studies on SMS patients [5].

The purpose of this brief chapter is to outline the principal clinical, laboratory and therapeutic features of "typical" stiff man syndrome and its "encephalomyelopathic" variants.

[1] Unit for Metabolic Diseases and Department of Medicine, San Raffaele Scientific Institute, Via Olgettina 60, 20132 Milan, Italy. e.mail: folli.franco@hsr.it
[2] Department of Neurology, Fondazione Mondino, University of Pavia, Italy

Aetiology and Pathogenesis

SMS Associated with Type 1 Diabetes Mellitus and Organ-Specific Autoimmune Diseases

The pathogenesis of SMS is a subject of active investigation. Clinical, pharmacological and laboratory evidence supports the concept that a functional impairment of the gamma-aminobutyric acid (GABA)-ergic inhibitory system may cause muscular rigidity and may be implicated in the increased prevalence of epilepsy, autonomic instability and the psychiatric symptoms of these patients [2, 6-17]. In 1986, the abrupt onset of diabetic ketoacidotic coma in a 48-year old woman affected by SMS and epilepsy fuelled the search for a possible pathogenic link between SMS and type 1 diabetes mellitus. To detect autoantibodies directed against the central nervous system, the serum and the cerebrospinal fluid of the patient were used to immunostain rat brain sections. The serum and the cerebrospinal fluid produced an identical, specific immunostaining of all brain regions examined [18]. The distribution of immunoreactivity corresponded to the known distribution of GABAergic nerve terminals [19-21], since the staining patterns produced were similar to those produced by sheep antibodies to glutamic acid decarboxylase (GAD), the enzyme responsible for the biosynthesis of GABA (Fig. 1a, c, e) [19, 20, 22, 23]. In double-immunofluorescence experiments, employing a sheep antiserum against GAD and the patient's serum/cerebrospinal fluid, the staining patterns were identical in all brain regions examined. SMS is associated with HLA phenotypes predisposing to type 1 diabetes and organ-specific autoimmunity (Table 1) [18, 24, 25].

Interestingly, outside the central nervous system, a high concentration of GAD and GABA has been found in pancreatic β-cells, male germ cells and oviduct and ovary [26-29]. Accordingly, the serum and cerebrospinal fluid of patients with SMS produced intense staining of rat pancreatic β-cells [3] (Fig. 1, panel G). Experiments conducted on a large series of patients affected by SMS demonstrated that the 65-kDa and 67-kDa isoforms of GAD were the autoantigens recognised by the majority (60%) of patients affected by SMS [30]. In this large series, it was confirmed that a high proportion of patients were affected by type 1 diabetes mellitus (30% in the anti-GAD-positive group) and that almost all patients

Table 1. Autoimmune diseases associated with SMS and SLS

Type I diabetes mellitus
Graves' disease
Hashimoto's thyroiditis
Pernicious anaemia
Vitiligo
Addison's disease
Alopecia totalis
Premature ovarian failure
Myasthenia gravis

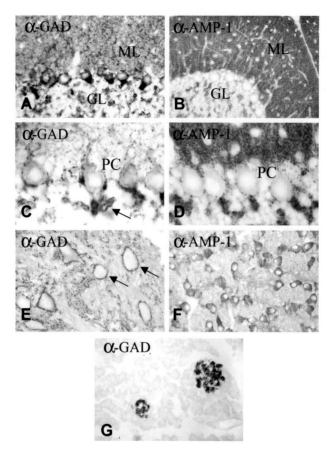

Fig. 1. A gallery of immunoperoxidase stainings of rat cerebellum, brain stem and pancreas with serum samples of a patient affected by stiff-man syndrome, type I diabetes mellitus and autoantibodies against glutamic acid decarboxylase (α-GAD; *panels A, C, E, G*) and of a patient affected by stiff-man syndrome, breast cancer and autoantibodies against amphiphysin I (α-AMP-1; *panels B, D, F*). Immunoreactivity is represented by the dark areas. *Panel A* shows a sagittal section of cerebellar cortex immunostained by a patient's serum containing autoantibodies against GAD. A very distinct pattern of immunoreactivity is visible in the molecular layer (*ML*) and in the granular layer (*GL*) of the cerebellum. *Panel C* shows a higher magnification of the Purkinje cell layer. There is intense immunoreactivity at the axon hillock (*arrow*) and immureactive dots outlining the perikaryon and the dendritic tree. *Panel E* shows a field in the deep cerebellar nuclei. Neuronal cell bodies are unstained and outlined by many immunoreactive dots, corresponding to GABAergic Purkinje cell axon terminals. *Panel G* shows a section of rat pancreas where the majority of the islets of Langerhans cells (insulin-producing b-cells) are heavily stained. *Panel B* shows a sagittal section of cerebellar cortex immunolabelled by a patient's serum with widespread staining of all regions containing synapses in both the molecular layer (*ML*) and the granular layer (*GL*). *Panel D* shows a higher magnification of the Purkinje cell layer, demonstrating that distinct dense staining corresponds to the distribution of synapses (all synapses). *Panel F* shows a deep cerebellar nuclei section with intense immunoreactivity of neuronal cytoplasm and dendrites as well as synaptic boutons on their surfaces. (Solimena M, Folli F, De Camilli P, unpublished material)

in the anti-GAD autoantibody positive group had clinical/serological evidence of organ-specific autoimmune diseases (Table 2) [30]. These findings have been confirmed in large series of patients by other investigators [31-35]. The demonstration that GAD is a major autoantigen in SMS, which is frequently associated with type I diabetes, also led, through collaborative studies, to the demonstration that the 64-kDa antigen of type 1 diabetes is in fact GAD [36]. This finding has allowed identification of individuals at risk of developing type 1 diabetes mellitus years before the development of clinical symptoms, and predict future insulin requirements in lean patients with type 2 diabetes mellitus (latent autoimmune diabetes of the adult, LADA) [37]. Autoantibodies directed against GAD are also found in late-onset cerebellar degeneration associated with endocrine disorders and sometimes with palatal myoclonus [38, 39] (see Chap. 8).

SMS Associated with Breast Cancer

In a few patients, SMS is associated with cancer, especially breast cancer [31, 40-46]. Immunocytochemistry experiments employing serum/cerebrospinal fluid from these patients demonstrated that the distribution of immunoreactivity is reminiscent of that of the synaptic vesicle proteins synapsin I and synaptophysin, which are synaptic-vesicle-associated proteins and are present in all synapses (Fig. 1B, D, F) [40, 47, 48]. The serum and cerebrospinal fluid of these patients recognized a nonintrinsic membrane protein of 128-kDa, which is expressed at high levels in the central nervous system and at lower levels in testis and endocrine tissues [40]. The 128-kDa autoantigen has been shown to be amphiphysin [49]. Autoantibodies directed against amphiphysin are also present in other paraneoplastic nervous system disorders (Table 2) [50-52].

Clinical Classification of SMS and Its Variants

"Classic" SMS

The "classic" SMS mainly affects adult, male and female patients. The progression of stiffness is slow (months); it involves axial muscles first, later extending to the proximal limb muscles. Painful muscle spasms are superimposed, and usually the stiffness and spasms are exaggerated by exteroceptive stimuli. Patients become hyperlordotic, with a hesitating gait and frequent falls. On palpation, the muscles have a rock-like consistency. Neurological examination shows no other abnormalities and intellect is not impaired. Patients who do not respond to proper symptomatic treatment may develop marked disability. SMS is frequently associated with type I diabetes (30%) or other polyendocrine-dysimmune diseases. Generalized epilepsy is also frequently seen, in around 10% of cases. Most patients respond to GABA-ergic drugs and have a positive response to immunosuppressive treatment such as intravenously immunoglobulin and plasma exchange (see below). Laboratory features include normal routine laboratory findings, except hypergly-

Table 2. Autoantibodies found in SMS, SLS and PERM

Anti-GAD antibodies (65-67 kDa isoforms) in association with type I diabetes mellitus
    polyendocrine autoimmune diseases
Anti-amphyphisin-I antibodies (128 kDa brain protein) in association with breast/lung/
    cancer or lymphoma cancer (*may be occult and should be searched for*)
Anti- gephyrin antibodies
Anti – 40/45 kDa brain protein band (*)
Anti-ICA 105/IA-2 antibodies(**)
Islet cell antibodies (ICA)
Anti-thyroid antibodies (thyroid microsomal fraction, thyroglobulin and TSH receptor)
Anti-parietal cells antibodies

Data from: * Solimena M. et al [30] NEJM; Folli F, Caradonna Z (2001) Unpublished observations
** Martino G et al [53]; Morgenthaler NG et al [54]

caemia in diabetics and sometimes slightly raised creatine kinase levels. Organ-
specific autoantibodies of polyendocrine syndrome are frequently detected along
with high titre serum and/or cerebrospinal fluid anti-GAD antibodies.

Routine CSF findings are usually normal and oligoclonal bands (OB) rarely
detected. Neuroimaging is normal. Needle or surface EMG recordings show
simultaneous CMUA at rest in both agonist and antagonist muscles, which is abolished by intravenously administered diazepam, peripheral nerve block, general
anaesthesia, or sleep. Exteroceptive reflexes are enhanced. The compound muscle
unit action potential in at least one axial muscle has been recently proposed as a
mandatory diagnostic criterion [1-3, 5]. The few pathological studies have shown
normal findings, except a recent one which suggested a loss of cerebellar
GABAergic neurons [55].

"Encephalomyelopathic" Variants of SMS

The encephalomyelopathic forms of SMS include three partially overlapping clinical variants: progressive encephalomyelitis with rigidity (PEMR) [56], jerking
SMS [57], and the stiff-limb syndrome (SLS) [58].

PERM is an acute or subacute illness with the worst prognosis (survival < 3
years), and is usually nonresponsive to various treatment strategies. Stiffness and
rigidity involve axial and limb muscles early, with painful spasms and myoclonus.
There are signs of brainstem and long-tract involvement. Cognitive impairment
is often prominent, with possible progression to stupor and coma. In the jerking
variant, limb myoclonus prevails and long-tract impairment is absent.

Jerking SMS is characterized by widespread rigidity, powerful and painful
spasms and reflex stimulus-sensitive disabling myoclonus mainly affecting axial
muscles, with good response to diazepam. In the originally reported patient, brain
CT showed cerebellar and brainstem atrophy. Prognosis is better (survival up to
10 years), with possible response to GABAergic drugs, immunosuppression and
plasma exchange.

SLS is characterised by prominent lower-limb stiffness and spasms, with occasional soft signs of central nervous system involvement. Survival is usually long-term, despite the fact that about one-half of patients become severely disabled. In this form, clinical-immunological relations are less striking: anti-GAD antibodies are rarely detected (15% of patients), and other antineuronal antibodies may be present [58-61]. A paraneoplastic aetiology is probable in many cases, especially in patients harbouring anti-amphiphysin antibodies [40, 49]. Inflammatory central nervous system changes, sometimes with the presence of OB may be found. A segmented CMUA has been a prevalent finding in SLS patients' limbs, in contrast to typical SMS patients, who mainly show a normal interference pattern [5, 58]. Patients with SLS may respond to GABAergic drugs, and some of them achieve remission with immunosuppressive treatment and, in paraneoplastic cases, tumour removal [5, 43, 58]. Few reported pathological studies support the hypothesis of an inflammatory-dysimmune pathogenesis, but agree with the nosological subdivision proposed by Brown and Marsden on the basis of the prominence of territorial involvement [5, 58].

Treatment

The therapeutic approach to the SMS complex remains rather heterogeneous. The small number of patients affected limits the possibilities of prospectively controlled studies, and the reported data are anecdotal. Nevertheless, some well-defined therapeutic possibilities are available: a) systemic symptomatic therapy; b) local symptomatic therapy; c) immunosuppressive therapy and d) plasmapheresis.

Systemic symptomatic therapy consists of long-term oral administration of GABAergic drugs [7-11, 62, 63]. Diazepam is the first choice (up to 100 mg/day). Baclofen (up to 60 mg/day), clonazepam (up to 6 mg/day), tizanidine (up to 6 mg/day), valproate (up to 2000 mg/day) and vigabatrin (up to 3000 mg/die) may be equally active. Response to treatment is usually better in typical SMS than in to SLS and PERM. Side effects include sedation, dysarthria, vertigo, ataxia and, sometimes, respiratory insufficiency.

Symptomatic local therapy consists of the intrathecal administration of baclofen [64] or intramuscular injection of botulinum toxin A (BTA) [65]. The former allows a four-fold increase in cerebrospinal fluid baclofen levels with only 1/100 of the usual oral dosage. A Medtronic or Infusaid pump is usually employed to administer a test bolus of 40-200 μg, followed by a starting dose of 50-240 μg/day reaching a maximum daily dose of 1600 μg. Patients with typical SMS respond better than PERM patients. Side effects are usually lower than with oral administration, but may include infection or catheter malfunction with a sudden reduction in drug supply, leading to severe relapse or to harmful overdose. Experience with BTA is limited, but good results have been seen in both SMS and SLS. Each upper or lower limb muscle is injected with 50-100 units of BTA, reaching a total amount of 700-1000 units. A positive response has been recorded up to

4 to 7 months after injection, allowing reduction in the use of systemic drugs and, sometimes, improvement in the contralateral muscle.

Immunosuppressive therapy consists in the administration of immunosuppressive drugs or intravenous immunoglobulin. We first reported remission of symptoms in two patients, one with typical SMS while the other had anti-amphiphysin antibodies and breast cancer with SLS features. Both had a poor response to systemic symptomatic drugs, showed steroid dependence and experienced long-lasting remission after tumour removal [42]. A good response has been reported in one patient with SLS features using high-dose intravenous 6-methylprednisolone (6MP) followed by intravenously administered cyclophosphamide bolus [66], and also in one PERM patient [67].

Intravenous administered immunoglobulins are usually given at a daily dosage of 400 mg/kg body weight for 5 consecutive days. By the end of one or more courses of this treatment, a positive result has been reported in SMS patients [68, 69] but not in an anti-amphipysin-positive SLS patient [66]. A controlled trial is now in progress [17].

Plasmapheresis was performed in the first anti-GAD autoantibody-positive SMS patient with good results, but other reports followed that gave conflicting results [70-74].

Typical SMS patients are often well controlled by systemic muscle relaxants; intrathecal baclofen may be a second choice. In non-responders we suggest intravenous immunoglobulin as a first immunoactive treatment and plasmapheresis as a second choice in diabetics. Non-diabetics may be first treated with steroids (oral administration or intravenous 6MP depending on severity, patient age, and

Table 3. Proposed diagnostic protocol in patients with SMS features

---

Anamnesis: prodromes, initial localisation of muscle aching or stiffness, progression modalities, known associated diseases with particular regard to endocrine-autoimmune diseases or neoplastic conditions

Neurological examination: neurological and mental status examination; localisation and degree of muscle stiffness; presence/absence of fixed trunk rigidity; presence / absence/ characteristics /precipitating factors of spasms and myoclonus; persistance/disappearance of rigidity during sleep

Response to GABAergic drugs (acute and/or chronic)

Electrophysiology: recording of continuous motor unit activity in limb and trunk muscles (before and after diazepam administration); exteroceptive reflexes study; cortical magnetic stimulation; multimodal evoked potential study

Laboratory and imaging: extensive laboratory analysis including organ-specific autoimmunity testing; routine cerebrospinal fluid and immunological studies; search for anti-neuronal antibodies in serum and cerebrospinal fluid (immunohistochemistry, western blotting, radioimmunoassay, ELISA)[a] ; brain and spinal cord MRI; MRI spectroscopy

---

[a] Screening for an occult malignancy in patients with anti-amphyphysin and other non anti-GAD antineuronal antibodies is recommended. Several commercial kits are available for the detection of anti-GAD or anti-amphyphisin autoantibodies

Table 4. Some distinctive clinical and laboratory features of stiff-man syndrome and "encephalomyelopathic" variants of stiff-man syndrome

	Stiff-man syndrome	Stiff-limb syndrome	Jerking Stiff-man syndrome	Progressive encephalomyelitis with rigidity
Trunk Stiffness	yes	yes/no	yes	yes
Limb Stiffness	yes/no	yes	yes	yes
Myoclonus	no	no	yes	yes/no
Continuous motor unit activity	yes (trunk)	yes/no	yes	yes
Response to Diazepam	yes	yes/no	yes/no	yes/no
Associated endocrine diseases	yes	yes/no	no	no
GAD autoantibodies	yes	yes/no	?	yes/no
Amphiphysin autoantibodies	yes/no	yes	?	yes/no

comorbidity). SLS patients may benefit from systemic or locally administered muscle relaxants and can be shifted to steroids if not controlled; intravenous immunoglobulin and plasmapheresis remain as further therapeutic choices. Tumour removal and treatment are indicated in paraneoplastic cases.

PERM represents a difficult therapeutic challenge. Due to its worse prognosis, a more aggressive treatment plan including muscle relaxants, various immuno-suppressive treatment modalities, intravenous immunoglobulin and plasma-pheresis should be considered. Supportive intensive care measures are sometimes indicated.

In conclusion, despite recent nosological efforts, the SMS complex cannot yet be subdivided in clear-cut clinical-immunological entities that could serve as a basis for strict guidelines for treatment. However, a reasonable diagnostic proto-col and treatment planning may be proposed on clinical and laboratory grounds (Tables 3 and 4).

Acknowledgements. Franco Folli is supported by Ministero della Sanità and Ministero Università e Ricerca Scientifica e Tecnologica (MURST). We thank Michele Solimena and Pietro De Camilli for their long and interesting collaboration, and Dr. Tara J. Zoll Folli for English editing.

References

1. Moersch FP, Woltmann HW (1956) Progressive fluctuating muscular rigidity and spasm (stiff-man syndrome): report of a case and some observations in 13 other cases. Mayo Clin Proc 31:421-427
2. Gordon EE, Januszko DM, Kaufman L (1967) A critical survey of stiff-man syndrome. Am J Med 42:582-599
3. Lorish TR, Thorsteinsson G, Howard FM (1989) Stiff-man syndrome updated. Mayo Clin Proc 64: 629-636
4. Piccolo G, Moglia A (1993) Stiff-man syndrome: one disease entity or more. In: Layzer RB, Sandrini G, Piccolo G, Martinelli M (eds) Motor Unit Hyperactivity States, Raven Press New York pp 23-30)
5. Brown P, Marsden CD (1999) The stiff man and stiff man plus syndrome. J Neurol 246:648-652
6. Howard FM Jr (1963) A new and effective drug in the treatment of stiff-man syndrome: preliminary report. Mayo Clin Proc 38:203-212
7. Cohen L (1966) Stiff-man syndrome: two patients treated with diazepam. JAMA 195:222-224
8. Westblom U (1977) Stiff-man syndrome and clonazepam. JAMA 237:1930
9. Spehlmann R, Norcross K, Rasmus SC, Schlageter NL (1981) Improvement of stiff-man syndrome with sodium valproate. Neurology 31:162-163
10. Meinck H-M, Conrad B (1986) Neuropharmacological investigations in the stiff-man syndrome. J Neurol 233:340-347
11. Boiardi A, Crenna P, Bussone G et al (1980) Neurological and pharmacological evaluation of a case of stiff-man syndrome. J Neurol 223:127-33
12. Henningsen P, Clement U, Kuchenhoff J et al (1996) Psychological factors in the diagnosis and pathogenesis of stiff-man syndrome. Neurology 47:38-42
13. Black JL, Barth EM, Williams DE, Tinsley JA (1998) Stiff-man syndrome. Results of interviews and psychologic testing. Psychosomatics 39:38-44
14. Tinsley JA, Barth EM, Black JL, Williams DE (1997) Psychiatric consultations in stiff-man syndrome. J Clin Psychiatry 58:444-449
15. Mitsumoto H, Schwartzman MJ, Estes ML et al (1991) Sudden death and paroxysmal autonomic dysfunction in stiff-man syndrome. J. Neurol 238:91-96
16. Floeter MK, Valls-Solè J, Toro C et al (1998) Physiologic studies of spinal inhibitory circuits in patients with stiff-person syndrome. Neurology 51:85-93
17. Levy LM, Dalakas MC, Floeter MK (1999) The stiff-person syndrome: an autoimmune disorder affecting neurotransmission of gamma-aminobutyric acid. Ann Intern Med 13:522-530
18. Solimena M, Folli F, Denis-Donini S et al (1988) Autoantibodies to glutamic acid decarboxilase in a patient with stiff-man syndrome, epilepsy, and type I diabetes mellitus. N Engl J Med 318:1012-1020
19. Oertel WH, Schmechel DE, Mugnaini E et al (1981) Immunocytochemical localization of glutamate decarboxylase in rat cerebellum with a new antiserum. Neuroscience 6:2715-2735
20. Mugnaini E, Oertel WH (1985) An atlas of the distribution of GABAergic neurons and terminals in the rat CNS as revealed by GAD immunohistochemistry. In: Björklund A, Hökfelt T (eds) Handbook of Chemical Neuroanatomy. Vol 4. GABA and neuropeptides in the CNS, Part I. Amsterdam, Elsevier Science Publishers, pp 436-608
21. Palay SL, Chan-Palay V (1974) Cerebellar cortex: cytology and organization. Springer-Verlag, New York

22. Kaufman DL, McGinnis JF, Krieger NR, Tobin AJ (1986) Brain glutamate decarboxilase cloned in labda γt-11: fusion protein produces gamma-aminobutyric acid. Science 232:1138-1140

23. Erlander MG, Tillakatne NJK, Feldblum S et al (1991) Two genes encode distinct glutamate decarboxilases. Neuron 7:91-100

24. Pugliese A, Gianani R, Eisenbarth GS et al (1994) Genetics of susceptibility and resistance to insulin-dependent diabetes in stiff-man syndrome. Lancet 344:1027-8

25. Williams AC, Nutt JG, Hare T (1988) Autoimmunity in stiff-man syndrome. Lancet 2:222

26. Reetz A, Solimena M, Matteoli M et al (1991) GABA and pancreatic β-cells: colocalization of glutamic acid decarboxilase (GAD) and GABA with synaptic-like microvesicles suggests their role in GABA storage and secretion. EMBO J 10:1275-1284

27. Vincent SR, Hokfelt T, Wu J-Y et al (1983) Immunohistochemical studies of the GABA system in the pancreas. Neuroendocrinology 36:197-204

28. Persson H, Pelto-Huikko M, Metsis M et al (1990) Expression of the neurotransmitter-synthesizing enzyme glutamic acid decarboxilase in male germ cells. Mol Cell Biol 10:4701-4711

29. Erdo SL, Joo F, Wolff JR (1989) Immunohistochemical localization of glutamate decarboxilase in the rat oviduct and ovary: further evidence for non-neural GABA systems. Cell Tissue Res 255:431-434

30. Solimena M, Folli F, Aparisi R et al (1990) Autoantibodies to GABA-ergic neurons and pancreatic beta cells in stiff-man syndrome. N Engl J Med (1990) 322:1555-1560

31. Grimaldi LME, Martino G, Braghi S et al (1993) Heterogeneity of autoantibodies in stiff-man syndrome. Ann Neurol 34:57-64

32. McEvoy KM (1991) Stiff-man syndrome. Seminars in Neurology 11:197-205

33. Toro C, Jacobowitz DM, Hallett M (1994) Stiff-man syndrome. Seminars in Neurology 14:154-158

34. Gorin F, Baldwin B, Tait R et al (1990) Stiff-man syndrome: a GABAergic autoimmune disorder with autoantigenic heterogeneity. Ann Neurol 28:71-74

35. Harding AE, Thompson PD, Kocen RS et al (1989) Plasma exchange and immunosuppression in the stiff man syndrome. Lancet 2:915

36. Baekkeskov S, Aanstoot H-J, Christgau S et al (1990) Identification of the 64K autoantigen in insulin-dependent diabetes as the GABA-synthesizing enzyme glutamic acid decarboxilase. Nature 347:151-156

37. Lernmark A (1996) Glutamic acid decarboxylase-gene to antigen to disease. J Intern Med 240:259-277

38. Saiz A, Arpa J, Sagasta A et al (1997) Autoantibodies to glutamic acid decarboxilase in three patients with cerebellar ataxia, late-onset insulin dependent diabetes mellitus, and polyendocrine autoimmunity. Neurology 49:1026-1030

39. Nemni R, Braghi S, Natali-Sora MG et al (1994) Autoantibodies to glutamic acid decarboxylase in palatal myoclonus and epilepsy. Ann Neurol 36:665-7

40. Folli F, Solimena M, Cofiell R et al (1993) Autoantibodies to a 128-kd protein in three women with the stiff-man syndrome and breast cancer. N Engl J Med 328:546-551

41. Masson C, Prier S, Benoit C et al (1987) Amnésie, syndrome de l' homme raide: Manifestations revelatrices d'une encéphalomyélite paranéoplastique. Ann Med Interne 138:502-505

42. Piccolo G, Cosi V, Zandrini C, Moglia A (1988) Steroid-responsive and dependent stiff-man syndrome: a clinical and electrophysiological study of two cases. Ital J Neurol Sci 9:559-566

43. Piccolo G, Cosi V (1989) Stiff man syndrome, dysimmune disorder, and cancer. Ann Neurol 26:105

44. Bateman DE, Weller RO, Kennedy P (1990) Stiff man syndrome: a rare paraneoplastic disorder? J Neurol Neurosurg Psichiatry 53:695-696

45. Silverman IE (1999) Paraneoplastic stiff limb syndrome. J Neurol Neurosurg Psychiatry 67:126-7

46. Valldeoriola F (1999) Movement disorders of autoimmune origin. J Neurol 246:423-431

47. De Camilli P, Cameron R, Greengard P (1983) Synapsin I (protein I), a nerve terminal-specific phosphoprotein. I. Its general distribution in synapses of the central and peripheral nervous system demonstrated by immunofluorescence in frozen and plastic sections. J Cell Biol 96:1337-1354

48. Navone F, Jahn R, Di Gioia G et al (1986) Protein p38: an integral membrane protein specific for small vescicles of neurons and neuroendocrine cells. J Cell Biol 103:2511-2527

49. De Camilli P, Thomas A, Cofiell R et al (1993) The synaptic vesicle-associated protein amphiphysin is the 128-kD autoantigen of Stiff-Man Syndrome with breast cancer. J Exp Med 178:2219-2223

50. Dropcho EJ (1996) Antiamphiphysin antibodies with small-cell lung carcinoma and paraneoplastic encephalomyelitis. Ann Neurol 43:659-667

51. Saiz A, Dalmau J, Butler MH et al (1999) Anti-amphiphysin I antibodies in patients with paraneoplastic neurological disorders associated with small cell lung carcinoma. J Neurol Neurosurg Psychiatry 66:214-7

52. Antoine JC, Absi L, Honnorat J et al (1999) Antiamphiphysin antibodies are associated with various paraneoplastic neurological syndromes and tumors. Arch Neurol 56:172-7

53. Martino G, Grimaldi LM, Bazzigaluppi E et al (1996) The insulin-dependent diabetes mellitus-associated ICA 105 autoantigen in stiff-man syndrome patients. J Neuroimmunol 69:129-134

54. Morgenthaler NG, Geissler J, Achenbach P et al (1997) Antibodies to the tyrosine phosphatase-like protein IA-2 are highly associated with IDDM, but not with autoimmune endocrine diseases or stiff man syndrome. Autoimmunity 25:203-211

55. Warich-Kirches M, Von Bossanyi P, Treuheit T et al (1997) Stiff-man syndrome: possible autoimmune etiology targeted against GABA-ergic cells. Clinical Neuropathology 16:214-219

56. Whiteley AM, Swash M, Urich H (1976) Progressive encephalomyelitis with rigidity. Its relation to "subacute myoclonic spinal neuronitis" and to the "stiff-man syndrome." Brain 99:27-42

57. Leigh PN, Rothwell JC, Traub M, Marsden CD (1980) A patient with reflex myoclonus and muscle rigidity: jerking stiff-man syndrome. J. Neurol Neurosurg Psichiatry 43:1125-1131

58. Barker RA, Revesz T, Thom M et al (1998) Review of 23 patients affected by stiff-man syndrome: clinical subdivision into stiff trunk (man) syndrome, stiff-limb syndrome, and progressive encephalomyelitis with rigidity. J Neurol Neurosurg Psychiatry 65:633-640

59. Folli F (1998) Stiff man syndrome, 40 years later. J Neurol Neurosurg Psychiatry 65:618

60. Saiz A, Graus F, Valldeoriola F et al (1998) Stiff-leg syndrome: a focal form of stiff-man syndrome. Ann Neurol 43:400-401

61. Shaw PJ (1999) Stiff-man syndrome and its variants. Lancet 353:86-87

62. Liguori R, Medori R, Marcello L et al (1993) Vigabatrin improves rigidity in stiff-person syndrome (SPS). Neurology 43:311

63. IA Sharoqi (1998) Improvement of stiff-man syndrome with vigabatrin. Neurology 50:833-834

64. Stayer C, Tronnier V, Dressnandt J et al (1997) Intrathecal baclofen therapy for stiff-man syndrome and progressive encephalomyelopathy with rigidity and myoclonus. Neurology 49:1591-1597

65. Liguori L, Cordivari C, Lugaresi E, Montagna P (1997) Botulinum Toxin A improves muscle spasms and rigidity in Stiff-person syndrome. Movement Disorders 12:1060-1063

66. Shmierer K, Valduzca JM, Bender A et al (1998) Atypical stiff-person syndrome with spinal MRI findings, amphiphysin autoantibodies, and immunosuppression. Neurology 51:250-252

67. McCombe PA, Chalk JB, Searle JW et al (1989) Progressive encephalomyelitis with rigidity: a case report with magnetic resonance imaging findings. J Neurol Neurosurg Psychiatry 52:1429-1431

68. Barker RA, Marsden CD (1997) Successful treatment of stiff-man syndrome with intravenous immunoglobulin. J Neurol Neurosurg Psychiatry 62:426-7

69. Amato AA, Cornman EW, Kissel JT (1994) Treatment of stiff-man syndrome with intravenous immunoglobulin. Neurology 44:1652-1660

70. Vicari AM, Folli F, Pozza G et al (1989) Plasmapheresis in the treatment of Stiff-Man Syndrome. N Engl J Med 320:1499

71. Hummel M, Durinovic-Bello I, Bonifacio E et al (1998) Humoral and cellular immune parameters before and after immunosuppressive therapy of a patient with stiff-man syndrome and insulin dependent diabetes mellitus. J Neurol Neurosurgery and Psychiatry 65:204-208

72. Hao W, Davis C, Hirsch IB et al (1999) Plasmapheresis and immunosuppression in stiff-man syndrome with type 1 diabetes: a 2-year study. J Neurol 246:731-735

73. Brashear HR, Phillips LH (1991) Autoantibodies to GABAergic neurons and response to plasmapheresis in stiff-man syndrome. Neurology 41:1588-1592

74. Anonimous (1996) Assessment of plasmapheresis. Report of the Therapeutics and Technology Assessment Subcommittee of the American Academy of Neurology. Neurology 47:840-843

Chapter 10

Paraneoplastic Neuropathies: a Neuropathological Overview

A. QUATTRINI

Neuropathies in patients with cancer are due to the direct effect of nerve compression or infiltration by tumour, or may occur as an indirect consequence of metabolic and nutritional disorders, infections, vascular disorders, or as a result of neurotoxic chemotherapy such as with vincristine, cisplatin and taxols [1, 2].

Paraneoplastic peripheral neuropathies (PPN) represent a "remote effect" of cancer on the peripheral nervous system. These are disorders of the peripheral nerve or its primary neuron that are caused by an indirect effect of carcinomas, mainly small-cell lung carcinomas (SCLC), lymphoproliferative disorders and dysproteinaemias. PPN may occur during the course of the tumour, but more usually present as the tumour's first manifestation and lead to its identification. PPN occur in a small percentage of patients with malignant neoplasms and they are frequently, though not always, associated with the presence of specific serum autoantibodies, which may be used as specific diagnostic aid. The pathological alterations in peripheral nerve are non-specific. In general, most of the peripheral neuropathies are of the axonal type and demyelination is uncommon. However, segmental demyelination may be found in some patients in association with axonal damage or with antibodies against myelin-associated glycoprotein. Finally, a few patients develop a multifocal mononeuritis multiplex in which the nerve is damaged indirectly by vasculitis that causes nerve ischeamia.

In this chapter we will discuss the specific paraneoplastic peripheral neuropathies, and also discuss the neuropathy associated with POEMS (polineuropathy, organomegaly, endocrinopathy, monoclonal gammopathy and skin changes) syndrome.

Specific Disorders

Subacute Sensory Neuronopathy

Subacute sensory neuronopathy (SSN), initially described by Denny-Brown [3], is recognised as a distinctive clinical syndrome and will be discussed first. SSN is

Department of Neuroscience, San Raffaele Scientific Institute, Via Olgettina 60, 20132 Milan, Italy. e-mail: a.quattrini@hsr.it characterised pathologically by a loss of neurons from dorsal root ganglia (DRG) [4]. A distinctive nuclear staining pattern of the Purkinje cells has been noted with sera derived from patients with SSN and an associated cancer, most commonly a SCLC. This antibody, anti-Hu or anti-neuronal nuclear antibody 1 (ANNA-1), has been reported in more than 80% of patients with paraneoplastic encephalomyelitis and/or SSN [5,6]. Patients develop a distal sensory disturbance with subacute numbness, dysaesthesia and paraesthesiae, and there may be associated features such as autonomic dysfunction, cerebellar ataxia or limbic encephalitis (see Chap. 7 for further information about the central nervous system disorders).

The main pathological findings are degeneration of neurons in the DRG associated with inflammatory infiltrates. Secondary changes in posterior nerve roots, peripheral sensory nerves and posterior columns of the spinal cord are observed. Usually, motor nerves are unaffected. The ganglia are damaged and inflammatory infiltrates such as T lymphocytes, predominantly cytotoxic T lymphocytes, and macrophages are found in the DRG. Inflammation may extend to posterior roots. The presence of inflammation depends on the duration of the disorder, disappearing with time. In post-mortem studies, another finding is the presence in the DRG of IgG deposits in sensory neurons and around the neuron [7]. Large lumbar sensory neurons are preferentially affected [8]. With the progression of the disorders, sensory neurons are lost and replaced by a proliferation of satellite cells (nodules of Nageotte). Involvement of the autonomic nervous system has also been observed. Neuronal cell degeneration in the sympathetic ganglia may be associated by inflammatory infiltrates.

Sural nerve biopsy shows non-specific findings. Mild to near total reduction of myelinated nerve fibres is observed with signs of active axonal degeneration (Fig. 1). Electron microscopy studies reveal that unmyelinated fibres can also be damaged. Axonal regeneration is not present or minimal, and segmental demyelination is not a prominent feature in SSN. However, pathological and immunohistochemical studies of peripheral nerves showed demyelination and remyelination

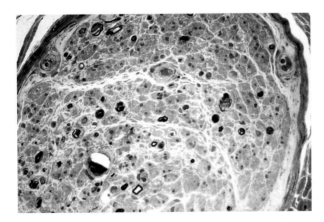

Fig. 1. Transverse semithin section of sural nerve from patients with subacute sensory neuropathy showing nonspecific findings: reduction of myelinated nerve fibres and signs of active axonal degeneration

with inflammatory CD8 T cells and macrophages in association with anti-Hu antibodies and SSN [9]. In many patients, the involvement of the peripheral nervous system is associated with paraneoplastic encephalomyelitis. In the central nervous system, in fact, the pathological process is a poly-encephalomyelitis with extension to cerebral and cerebellar white matter [10].

Patients with SSN should not undergo sural nerve biopsy, as biopsy specimen examination is non-specific. Although the pathogenic role of specific antineuronal antibodies has not yet been established, anti-Hu antibodies have a high diagnostic value for paraneoplastic SSN.

Motor Neuron Disease

The relationship between motor neuron disease (MND) and cancer is debatable [11]. MND is observed in patients with lung or renal tumours, thymoma and lymphoma. Although epidemiological studies have not demonstrated an increased incidence of cancer in patients with MND, paraneoplastic disorders may be associated with MND. The following evidence suggests that some forms of MND may be paraneoplastic: (1) treatment of associated cancer clearly improves the neurological disorder; (2) the presence in the patients' sera of anti-Hu antibodies; (3) the frequent association of lymphoproliferative disorders with subacute motor neuropathy [12-15].

Post-mortem studies in these patients showed severe neuronal degeneration and loss of anterior horn cells with the presence of perivascular inflammatory infiltrates in the spinal cord grey matter. Lymphoid infiltrates surrounding residual neurons and pronounced gliosis were also observed [12, 13]. In one patient with SCLC and MND, high titres of anti-Hu antibodies were detected [12]. In some cases, demyelination has been observed in the white matter of spinal cord, particularly that surrounding the grey matter columns.

MND and subacute motor neuropathy have been reported in patients with lymphoma [14, 15]. In the series of patients described by Younger et al. [15], paraproteinaemia was frequently detected, suggesting that an immunological disorder may play a role in the pathogenesis of MND. Pathological studies showed marked loss of motor neurons in spinal cord, brain stem and motor cortex without inflammatory infiltrates. Degeneration of the corticospinal tract was also observed. Sural nerve biopsy examination showed axonal degeneration and scattered inflammatory cell infiltrates in one patient and demyelination in another case.

Sensorimotor Neuropathies

Sensorimotor neuropathies have been associated with underlying cancer such as lung cancer, breast and ovarian cancers, lymphoma and multiple myeloma [16]. These neuropathies are a relatively frequent complication of cancer and in their clinical course may be acute, mimicking Guillain-Barré syndrome or relapsing and remitting neuropathies, and/or subacute or chronic sensorimotor neuropathies.

The pathological features associated with sensorimotor neuropathies are heterogeneous. Demyelination has been reported in haematological cases with features of a typical macrophage-mediated demyelination [17], but axonal damage without inflammatory infiltrates has also been observed [18]. Onion bulb formation may also be found. Although nerve biopsy does not shown any specifically histological features in the majority of patients, a search for an occult neoplasm is indicated in some instances when no other cause of the neuropathy can be determined. However, in some patients with lymphomas and sensorimotor neuropathy, the sural nerve biopsy is specific in showing direct infiltration of the nerve by the tumour cells.

Inflammation of peripheral nerve is not a typical feature. However, perivascular inflammation and vasculitis have been reported in patients with sensorimotor multineuropathy. Such cases seem to reflect disease of the peripheral nerve rather than the spinal ganglion. The diagnosis must be confirmed by the demonstration of vasculitis in the nerve biopsy. It is important to show nerve vasculitis because it is a rare but treatable condition [19-21].

POEMS Syndrome

POEMS syndrome is a multisystemic disorder characterised by polyneuropathy, organomegaly, endocrinopathy, monoclonal gammopathy and skin changes [22, 23]. Many of these patients have osteosclerotic bone lesions. The pathogenesis of these multiple systemic manifestations is still unknown. Peripheral neuropathy is a nearly universal finding in POEMS syndrome. The neuropathy is usually a progressive distal sensorimotor polyneuropathy [24]. The peripheral nerve pathology is variable and includes axonal degeneration and demyelination [24, 25]. Endoneurial oedema, another prominent feature of this peripheral neuropathy, and microvascular abnormalities [26] that may be also present, could account for some systemic manifestations of the syndrome. Gammaglobulin deposition in the endoneurial interstitium [27] may also relate to the neural dysfunction.

Case Report. A 52-year-old woman with POEMS syndrome was evaluated in June 1998. A plasmocytoma was diagnosed pathologically in 1992, when a solitary osteosclerotic lesion was found on the seventh cervical vertebral body and removed. Radiotherapy was then given. In 1996, the patient developed a sensorimotor demyelinating polyneuropathy, hepatosplenomegaly, adrenal insufficiency and hypothyroidism, serum monoclonal gammopathy of IgG λ type, skin hyperpigmentation and sclerodactyly, and was diagnosed as having a POEMS syndrome. In February 1997, a bone marrow biopsy revealed the presence of neoplastic plasma cells and polychemotherapy was started. In November 1997 a control bone marrow biopsy showed a normal plasma cell number, but sensorimotor functions deteriorated further.

During hospitalisation, the patient underwent general and neurological evaluations. Hepatosplenomegaly and generalised skin hyperpigmentation and thickness were noted.

Neurological examination showed symmetrical reduction of strength (grade

3/5 MRC) in the distal muscles of the arms and legs. Tendon reflexes were absent. Flexor plantar reflexes were present bilaterally. Touch, temperature, pinprick, joint position and vibration sensation were diminished in a stocking-glove distribution. The patient was ataxic with a widebased gait. Romberg sign was present.

Electrophysiological investigations revealed a denervation pattern in tested muscles, and nerve conduction studies showed diffuse sensorimotor polyneuropathy, primarily demyelinating with axonal features. Motor conduction velocities were decreased in the median nerve and in the peroneal nerve and no response of the sural nerve was found. Motor action potentials were decreased in the peroneal and median nerves; sensory action potentials of sural nerves were absent. Distal motor latencies were increased.

The patient became progressively weaker and her neurological condition deteriorated, with anaesthesia and reduced strength (grade 2/5 MRC) in all limbs. At this time, repeated tests of neurophysiological parameters worsened. A sural nerve biopsy was performed proximal to the lateral malleolus under informed consent. Specimens were examined by light and electron microscopy. By direct immunoperoxidase staining, the endoneurium was strongly positive for IgG and λ light chains. Semithin sections showed loss of myelinated nerve fibres and pronounced endoneurial oedema (Fig. 2A). Prominent thickening of endoneurial microvessel basement membranes was present (Fig. 2B, 2C). Infiltrations of inflammatory and red blood cells in the subendothelial space were occasionally observed (Fig. 2C). Electron microscopy confirmed axonal damage, demyelina-

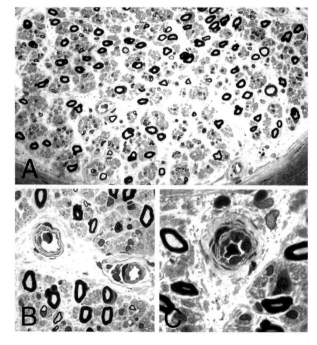

Fig. 2 A-C. Transverse semithin sections of sural nerve from a patient with POEMS syndrome showing generalised reduction of myelinated nerve fibres associated with prominent endoneurial oedema (**A**). The vessels show abnormally thickened walls (**B, C**). A few red blood and inflammatory cells are present in the perivascular space (**C**)

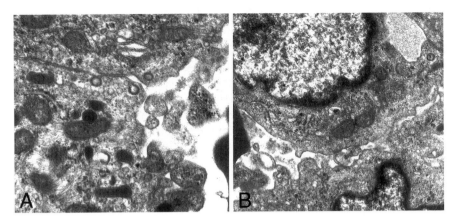

Fig. 3 A, B. Electron micrographs showing endothelial cytoplasmic enlargement. **A** Inclusions, microfilaments and many pinocytic vesicles adjacent to the cell membranes are present. **B** shows gap between endothelial cells and subendothelial amorphous material

tion, aberrant proliferative basement membrane with reduplication around endothelial cells and pericytes, and amorphous material in subendothelial space. Many endothelial cells were very swollen with an organelle-rich cytoplasm. Accumulation of intracytoplasmic filaments, inclusions and pinocytotic vesicles adjacent to the cell membranes were observed (Fig. 3A). Some of the tight junctions between endothelial cells of endoneurial microvessels had disappeared, and obvious gaps were found between endothelial cells (Fig. 3B). No amyloid deposits were observed.

Treatment with prednisone (1 mg/kg per day for 10 days) was begun, then slowly tapered within 2 weeks and maintained at 0.25 mg/kg per day for two months. The patient reported prompt improvement of neurological symptoms as well as of skin abnormalities a few weeks after the beginning of therapy. Re-evaluation 2 months later revealed improvement of peripheral sensory and motor symptoms and signs. Repeated neurophysiological studies showed a reappearance of sural sensory action potential, peroneal motor action potentials and reduction of distal motor latencies mainly of the median nerve.

In this case microscopic evaluation of peripheral nerve showed marked alterations of endothelial cells, marked oedema and IgG λ deposits in the endoneurium. These findings suggest an increase in capillary permeability. The blood-peripheral nerve barrier is not as tight as the blood-brain barrier, so small amounts of circulating proteins, such as IgG, which cannot enter the central nervous system, can enter the endoneurial space. Barrier dysfunction at the endothelial level may play a role in the pathophysiology of this disease and may result in widespread oedema of optic and peripheral nerves as well as, perhaps, other organs. Others have suggested that microvascular abnormalities could play a role in POEMS syndrome with generalised endothelial injury [26, 28]. The monoclonal paraprotein or another abnormal tumour product could act as a "permeability" factor exerting direct damage or causing functional changes of the vascular endothelium [29, 30].

As was the case in our patient, corticosteroids may be beneficial when peripheral nerve oedema is present. Discovery of endoneurial oedema on nerve biopsy might suggest a potential for corticosteroid responsiveness of the peripheral nerve. It is likely that oedema plays an important role in the progression of nerve dysfunction by altering the nerve microenvironment. In conclusion, our results suggest that steroid-responsive microvascular abnormalities could play a pathogenic role in the nervous system manifestations of POEMS syndrome.

References

1. Hughes R, Sharrack B, Rubens R (1996) Carcinoma and the peripheral nervous system. J Neurol 243:371-376
2. Fazio R, Quattrini A, Bolognesi A et al (1999) Docetaxel neuropathy: a distal axonopathy. Acta Neuropathol 98:651-653
3. Denny-Brown D (1948) Primary sensory neuropathy with muscular changes associated with carcinoma. J Neurol Neurosurg Psychiatry 11:73-87
4. Horwich MS, Cho L, Porro S, Posner JB (1997) Subacute sensory neuronopathy: a remote effect of carcinoma. Ann Neurol 2:7-19
5. Oh SJ, Dropco EJ, Claussen GC (1997) Anti-Hu-associated paraneoplastic sensory neuronopathy responding to early aggressive immunotherapy: report of two cases and review of the literature. Muscle Nerve 20:1576-1582
6. Graus YF, Verschuuren JJ, Degenhardt A et al (1998) Selection of recombinant anti-HuD Fab fragments from a phage display antibody library of a lung cancer patient with paraneoplastic encephalomyelitis. J Neuroimmunol 82:200-209
7. Graus F, Ribalta T, Campo E et al (1990) Immohistochemical analysis of the immune reaction in the nervous system in paraneoplastic encephalomyelitis. Neurology 40:219-222
8. Ohnishi A, Ogawa M (1986) Preferential loss of large lumbar sensory neurons in carcinomatous sensory neuropathy. Ann Neurol 20:102-104
9. Antoine JC, Mosnier JF, Honnorat J et al (1998) Paraneoplastic demyelinating neuropathy, subacute sensory neuropathy and anti-HU antibodies: a clinicopathological study of an autopsy case. Muscle Nerve 21:850-857.
10. Scaravilli F, Shu F, Groves M, Thom M (1999) The neuropathology of paraneoplastic syndromes. Brain Pathol 9:251-260
11. Rowland LP (1997) Paraneoplastic primary lateral sclerosis and amyotrophic lateral sclerosis. Ann Neurol 41:703-705
12. Verma A, Berger JR, Snodgrass S, Petito C (1996) Motor neuron disease: a paraneoplastic process associated with anti-Hu antibody and small-cell lung carcinoma. Ann Neurol 40:112-116
13. Forsyth PA, Dalmau J, Graus F et al (1997) Motor neuron syndromes in cancer patients. Ann Neurol 41:722-730
14. Shold SC, Cho ES, Somasundaram M, Posner JB (1979) Subacute motor neuropathy: a remote effect of lymphoma. Ann Neurol 5:271-287
15. Younger DS, Rowland LP, Latov N et al (1991) Lymphoma, motor neuron diseases, and amyotrophic lateral sclerosis. Ann Neurol 29:78-86

16. Croft PB, Urich H, Wilkinson M (1967) Peripheral neuropathy of sensorimotor type associated with malignant disease. Brain 90:31-65)

17. Julien J, Vital C, Aupy G (1980) Guillain-Barré syndrome and Hodgkin's disease. Ultrastructural study of a peripheral nerve. J Neurol Sci 45:23-27

18. Schlaepfer WW (1974) Axonal degeneration in the sural nerves of cancer patients. Cancer 34:371-381

19. Vincent D, Dubas F, Hauw JJ et al (1986) Nerve and muscle microvasculitis in peripheral neuropathy: a remote effect of cancer. J Neurol Neurosurg Psychiatry 49:1007-1010

20. Oh SJ, Slaughter R, Harrell L (1991) Paraneoplastic vasculitic neuropathy: a treatable neuropathy. Muscle Nerve 14:152-156

21. Said G (1997) Necrotizing peripheral nerve vasculitis. Neurol Clin 15:836-838.

22. Miralles GD, O'Fallon JR, Talley NJ (1992) Plasma-cell dyscrasia with polyneuropathy. The spectrum of POEMS syndrome. N Eng J Med 327:1919-1923

23. Bolling JP, Brazis PW (1990) Optic disk swelling with peripheral neuropathy, organomegaly, endocrinopathy, monoclonal gammopathy, and skin changes (POEMS syndrome). Am J Pathol 109:503-510

24. Nakanishi T, Sobue I, Toyokura Y (1984) The Crow-Fukase syndrome: a study of 102 cases in Japan. Neurology 34:712-720

25. Donaghy M, Hall P, Gawler J (1989) Peripheral neuropathy associated with Castleman's disease. J Neurol Sci 89:253-267

26. Saida K, Kawakami H, Ohta M, Iwamura K (1997) Coagulation and vascular abnormalities in Crow-Fukase syndrome. Muscle Nerve 20:486-492

27. Adams D, Said G (1998) Ultrastructural characterisation of the M protein in nerve biopsy of patients with POEMS syndrome. J Neurol Neurosurg Psychiatry 64:809-812

28. Soubrier MJ, Dubost JJ, Sauvezie BJM (1994) POEMS syndrome: a study of 25 cases and a review of the literature. Am J Med 97:543-553

29. Watanabe O, Maruyama I, Arimura K et al (1998) Overproduction of vascular endothelial growth factor/vascular permeability factor is causative in Crow-Fukase (POEMS) syndrome. Muscle Nerve 21:1390-1397

30. Gherardi RK, Belec L, Soubrier M, et al (1996) Overproduction of proinflammatory cytokines imbalanced by their antagonists in POEMS syndrome. Blood 87:1458-1465

Chapter 11

Opsoclonus-Myoclonus Syndrome in Childhood

F. Blaes[1], B. Lang[2]

Introduction

In 1962 Kinsbourne described six infants with a syndrome of myoclonus and "dancing eyes" (opsoclonus-myoclonus syndrome, OMS) [1]. OMS in association with neuroblastoma [2] and other neural-crest-derived tumours has been described in some children (paraneoplastic OMS, OMS-NB[+]), whilst several other potential causal relationships have also been found (Table 1), but little progress has been made in understanding the etiology and treatment of this syndrome. Recently, however, there has been increasing evidence that OMS may be an autoimmune disease of the central nervous system. This chapter will describe the clinical and immunological features as well as treatment options of OMS.

Clinical Features

OMS in children typically starts within the first 3 years of life, often around the age of 18 months, with a subacute onset of chaotic, synchronous eye movements (opsoclonus) and spontaneous jerks of the limb and face muscles (myoclonus). The opsoclonus, described as multi-directional conjugate saccades, is often provoked by fixation and usually persists during sleep. The frequency of the oscillation is irregular, usually ranging from 6 to 12 Hz. Besides the opsoclonus-myoclonus, these children also develop ataxia, more evident in the trunk than in the limbs, with gait disturbance and unsteadiness [3]. Other symptoms, often described by the parents, include sleep disturbances and an excessive irritability in the children. Moreover, there is increasing evidence from long-term follow-up of children with OMS that they frequently have learning and behavioural difficulties. Preliminary data from long-term follow-up suggest that the learning difficulties persist over years and tend to worsen. In a retrospective study, 69% of paraneoplastic OMS children had persistent neurological deficits including speech delay, cognitive deficits, motor delay and behavioural problems [4].

[1] Department of Neurology, Justus-Liebig University, Giessen, Germany. e-mail: franz.blaes@neuro.med.uni-giessen.de

[2] Neurosciences Group, Weatherall Institute of Molecular Medicine, John Radcliffe Hospital, Oxford, United Kingdom

Table 1. Possible causes of opsoclonus-myoclonus syndrome

Viruses
    Mumps (paramyoviridae) [46]
    Epstein-Barr virus [47,48]
    Coxsackie B3 [49]
Neoplasms
    Neuroblastoma [2, 5, 6, 50, 51, 52]
    Hepatoblastoma [5]
    Ganglioneuroma [6]
Other agents known to cause symptoms of opsoclonus
    Toxic
        Thallium
        Strychnine
        Organophosphates
        Toluene
    Drugs
        Amytriptyline
        Lithium
        Phenytoin

A preceding infection, bacterial or viral, occurs in approximately 50% of OMS cases; however, no clear temporal relation has been established with immunisation. The percentage of OMS cases with associated tumours is still unclear. The tumours of children with OMS-NB⁺ are almost exclusively neuroblastomas or ganglioneuromas, although a few patients with other associated tumours have been reported [5]. Some authors report 2-10% of OMS cases to be associated with neuroblastoma [6]. However, recent data suggest that, as screening techniques improve, many more cases of OMS may be found to be associated with an underlying neuroblastoma. Interestingly, children who present with neuroblastoma and OMS have a better prognosis than those who present without the neurological condition [7]. The neuroblastomas associated with OMS appear to be small, well-differentiated tumours with a favourable histological subtype. These tumours express a low number of *n-myc* copies, a marker also known to be associated with a better prognosis [8].

Thus, in summary, OMS presents principally in early childhood and appears to exist in two forms, either with or without associated tumour. The suggestion that children with paraneoplastic OMS have a better prognosis than those without the neurological disease is highly reminiscent of observations in the Lambert-Eaton myasthenic syndrome, an autoimmune disorder of the peripheral neuromuscular junction which presents with and without small-cell lung cancer [9].

Pathology and Site of the Lesion

Pathological studies of OMS patients are very limited. A biopsy taken from the cerebellar vermis of a child with OMS revealed Purkinje cell and granular cell loss with gliosis [10], whilst in another paraneoplastic case of OMS, similar changes in the cerebellum have been found [11]. The destruction of Purkinje cells in other neurological diseases, such as cerebellar degeneration, does not usually cause opsoclonus or myoclonus. However, the definite site of the lesion causing the neurological symptoms in opsoclonus is still unknown. Other authors have suggested brainstem nerve cells, the "omnipause" neurons, as a possible site of the lesion in this disease [12]; however, induction of opsoclonus was not caused by destruction of these neurons in animal experiments [13].

Treatment

A wide range of treatments is currently used in OMS children, with different strategies being used in paraneoplastic and non-paraneoplastic cases. The immunosuppressive treatment of OMS has not changed very much since the original description of the disease by Kinsbourne in 1962 [1]. This includes the use of steroids and ACTH, both of which lead to a rapid improvement of the neurological symptoms [14]. However, there are no conclusive data showing a significant advantage of one of these drugs over the other, and in most patients withdrawal of steroid or ACTH therapy has been shown to cause frequent relapses, so therapy has to be continued for several years. The mechanisms by which ACTH and steroids influence the immune system in OMS are not exactly clear. ACTH seems to act as a direct immunosuppressive and not via release of corticosteroids, and appears to inhibit the antibody response to T-cell dependent antigens [15]. Steroids decrease lymphocytic differentiation and proliferation and inhibit phagocytosis. They also suppress production of several interleukins [16]. However, beside the immunosuppressive actions of these drugs, there are direct effects of steroids on organotypic cerebellar cell cultures (F. Blaes et al., unpublished results), which could additionally influence the symptoms of OMS. Immunomodulatory therapies have also been used. There are a few case reports and small studies of the use of plasmapheresis or protein A column immunoabsorption in the treatment of child hood OMS [17-19].

Intravenous immunoglobulins (IVIg) have also been reported to be effective in the initial treatment of OMS [20, 21]. Unfortunately, a single IVIg treatment does not lead to long-term improvement in most OMS children. The mechanism of action of IVIg has also not been established clearly. IVIg down-regulates interleukin-2 production by T cells [22], can influence a variety of cytokines and can increase the IgG catabolism in general (for review see [23]). In addition, IVIg preparations contain soluble MHC molecules [24] and soluble CD4 and CD8 molecules, all of which are able to down-regulate T cell activation [25]. In antibody-mediated diseases, a direct idiotype-anti-idiotype interaction by IVIg has also been considered a main mechanism of action [26]. However, at present there are insufficient data available about the special mechanisms of action in OMS.

If a tumour is found, it is usually removed. Theoretically, there may be a conflict in using immunosuppressive therapies in patients who are mounting a hypothetically sufficient anti-tumoral immune response [27]. There are no data, however, that indicate a progressive course of the tumour in OMS children receiving immunosuppressants. On the contrary, an interesting observation was made in a retrospective study of the long-term neurological outcome in neuroblastoma-associated OMS [4]. Six out of nine children who received chemotherapy as a part of their neuroblastoma treatment had complete recovery from the OMS, compared to only 2 out of 19 children who had surgery alone, indicating a possible immunosuppressive effect of the chemotherapeutic drugs.

Taken together, the most established therapies at present are steroids, ACTH and IVIg. Since no comparative data exist about these therapies, a multi-centre study should be undertaken to find out the most efficient treatment for these children. This is especially important in the context of the long-term neurological and neuropsychological disturbances that persist in most of these children.

Immunopathology

Immunological abnormalities in the CSF have been found in several OMS children. Levels of IgG and IgM in the CSF may be increased, and the presence of oligoclonal bands has been described [28, 29, 30, 31]. Moreover, an abnormal reactivity of lymphocytes to neuroblastoma extracts has been described [32]. In addition, a retrospective study has shown that children with neuroblastoma with a favourable outcome have lymphocytic infiltration of their tumours [33]. Conversely, Ollert et al. demonstrated the presence of natural cytotoxic IgM in the sera of healthy controls but not in the sera of patients with neuroblastoma [34]. This IgM recognises a 260-kDa neuroblastoma protein [35]. These data suggest that a natural immunity to the tumour, or the lack of it, might be important in determining susceptibility to neuroblastoma and the ability to raise an immune response.

The description of autoantibodies in children with OMS has strengthened the hypothesis of an autoimmune pathology. Several different autoantibodies have been described in the sera of OMS children both with and without neuroblastoma. Anti-neurofilament antibodies could be detected in the serum of children with OMS [36, 37]. However, these antibodies are also described in other neurological diseases and even healthy controls [38]. In some children with paraneoplastic OMS, anti-Hu antibodies have been described [21, 39]. These antibodies react with a group of 35- to 40-kDa RNA-binding proteins and are usually associated with small-cell lung cancer and paraneoplastic neurological syndromes in adults [40, 41] (see Chap. 7). By contrast, the anti-Ri antibody, which is associated with paraneoplastic OMS in adults [42], has not been found in children with OMS. Antunes et al. described a variety of immunohistochemistry and western blot reactivities in neuroblastoma-associated OMS, but the only common reactivity in their patients was anti-Hu antibody in patients with neuroblastoma with and without associated OMS [39]. Recently, Connolly and coworkers described autoreactivity against cytoplasmic Purkinje and granular cells and a pattern of common reactivities using western blot analysis in paraneoplastic and non-paraneoplastic OMS [37]. However, in our own experiments using similar techniques we were unable to detect a common reactivity pattern in various cerebellar and neuroblastoma protein fractions in ten OMS patients with and without an underlying tumour (E. Blaes et al., in preparation). Nevertheless, using immunohistochemical techniques we were able to detect cytoplasmic staining of the Purkinje cell in some non-paraneoplastic OMS patients. Moreover, we found that some of the OMS patients had IgG surface staining of neuroblastoma cells, and that preincubation of neuroblastoma cells in OMS sera (both paraneoplastic and non-paraneoplastic), obtained during active disease, resulted in growth inhibition of neuroblastoma cells in vitro (F. Blaes et al., in preparation). These experiments show the presence of serum autoantibodies, capable of cross-reacting with brain and neuroblastoma antigens, that have functional effects on neuronal cells.

As yet, there is no identified autoantigen in OMS. From a neuro-ophthalmological view, OMS may represent a form of overexcitation or disinhibition. One may speculate that the autoimmune reaction in OMS may involve either inhibitory neurotransmitters such as GABA or glycine, or excitatory transmitters such as glutamate. Interestingly, the Nova antigens recognised by the OMS-associated anti-Ri antibody in adults (see Chap. 7) are probably involved in the processing of glycine receptor mRNA [43]. Alternatively, the finding of low CSF levels of the dopamine metabolite homovanillic acid and the serotonin metabolite 5-hydroxyindoleacetic acid supports the hypothesis that the disorder is caused by a disturbance of the neurotransmission of one or both of these neurotransmitters [44]. However, there is no clear evidence for involvement of monoaminergic neuronal pathways in the development of eye movement disturbances or myoclonus.

Taken together, the presence of autoantibodies in OMS suggests an autoimmune pathogenesis of the disease. However, the lack of finding of a common autoantigen in the reported studies is disappointing. One possible explanation may be that the antigenic targets under suspicion are surface molecules present at low density and the binding cannot be detected with most standard antibody-screening methods; this is supported by our finding of surface binding against neuronal cells in some of these patients. To elucidate OMS as an antibody-mediated disease, a passive transfer of OMS-IgG in animals should be undertaken. An alternative explanation could be that the disease is T-cell-mediated. In adult paraneoplastic syndromes there is increasing evidence for a major role of cytotoxic T cells in the pathophysiology of these diseases [45], but there are very few data in the literature on the T cell responses in children.

In conclusion, there is increasing evidence that OMS with or without an associated tumour has an autoimmune aetiology and can be alleviated by immunomodulatory therapy. The target of the autoimmune attack is as yet unknown, and the elucidation of the target molecule(s) will provide important information on how the immune system and the nervous system interact.

References

1. Kinsbourne M (1962) Myoclonic encephalopathy of infants. J Neurol Neurosurg Psychiatry 25:271-276
2. Soloman GE, Chutorian AB (1968) Opsoclonus and occult neuroblastoma. N Engl J Med 279:475-477
3. Averbuch-Heller L, Remler B (1996) Opsoclonus. Semin Neurol 16:21-26
4. Russo C, Cohn SL, Petruzzi MJ, de Alarcon PA (1997) Long-term neurologic outcome in children with opsoclonus-myoclonus associated with neuroblastoma: a report from the pediatric oncology group. Med Ped Oncol 29:284-288
5. Wilfong AA, Parke JT, McCrary JA (1992) Opsoclonus-myoclonus with Beckwith-Wiedermann syndrome and hepatoblastoma. Pediatr Neurol 8:77-79
6. Pohl KR, Pritchard J, Wilson J (1996) Neurological sequelae of the dancing eye syndrome. Eur J Pediatr 155:237-244
7. Altman AJ, Baehner RL (1976) Favorable prognosis for survival in children with coincident opso-myoclonus and neuroblastoma. Cancer 37:846-852
8. Cohn SL, Salwen H, Herst CV et al (1988) Single copies of the n-myc oncogene in neuroblastomas from children presenting with the syndrome of opsoclonus-myoclonus. Cancer 62:723-726
9. Maddison P, Newsom-Davis J, Mills KR, Souhami RL (1988) Favourable prognosis in Lambert-Eaton myasthenic syndrome and small cell lung carcinoma. Lancet 353:117-118
10. Tuchman RF, Alvarez LA, Kantrowitz AB et al (1989) Opsoclonus-myoclonus syndrome: correlation of radiographic and pathological observations. Neuroradiology 31:250-252
11. Ziter FA, Bray PF, Cancilla PA (1979) Neuropathological findings in a patient with neuroblastoma and myoclonic encephalopathy. Arch Neurol 36:51
12. Ridley A, Kennard C, Scholtz CL et al (1987) Omnipause neurons in two cases of opsoclonus associated with oat cell carcinoma of the lung. Brain 110:1699-1709
13. Kaneko CR (1996) Effect of ibotenic acid lesions of the omnipause neurons on saccadic eye movements in rhesus macaques. J Neurophysiol 75:2229-2242
14. Pranzatelli MR (1996) The immunopharmacology of the opsoclonus-myoclonus syndrome. Clin Neuropharmacol 19:1-47
15. Johnson HM, Smith EM, Torres BA, Blalock JE (1982) Regulation of in vitro antibody response by neuroendocrine hormones. Proc Natl Acad Sci USA 79:4171-4174
16. Munck A, Guyre PM (1991) Glucocorticoids and immune function. In: Ader R, Felten DL, Cohen N (eds) Psychoneuroimmunology. Academic Press, San Diego, pp 447-472
17. Sugie H, Sugie Y, Akimoto H et al (1992) High-dose i.v. human immunoglobulin in a case with infantile opsoclonus polymyoclonia syndrome. Acta Paediatr 81:371-372
18. Nolte MT, Pirofsky B, Gerritz GA, Golding B (1979) Intravenous immunoglobulin therapy for antibody deficiency. Clin Exp Immunol 36:237-243
19. Cher LM, Hochberg FH, Teruya J et al (1995) Therapy for paraneoplastic neurologic syndromes in six patients with protein A column immunoadsorption. Cancer 75:1678-1683
20. Nitschke M, Hochberg F, Dropcho E (1995) Improvement of paraneoplastic opsoclonus-myoclonus after protein A column therapy. N Engl J Med 332:192
21. Fisher PG, Wechsler DS, Singer HS (1994) Anti-Hu antibody in a neuroblastoma-associated paraneoplastic syndrome. Pediatr Neurol 10:309-312
22. Modiano JF, Amran D, Lack G et al (1997) Posttranscriptional regulation of T-cell IL-2 production by human pooled immunoglobulin. Clin Immunol Immunopathol 83:77-85

23. Stangel M, Hartung HP, Marx P, Gold R (1998) Intravenous immunoglobulin treatment of neurological autoimmune diseases. J Neurol Sci 153:203-214
24. Grosse-Wilde H, Blasczyk A, Westhoff U (1992) Soluble HLA class I and class II concentrations in commercial immunoglobulin preparations. Tissue Antigens 39:74-77
25. Blasczyk R, Westhoff U, Grosse-Wilde H (1993) Soluble CD4, CD8, and HLA molecules in commercial immunoglobulin preparations. Lancet 341:789-790
26. Ronda N, Haury M, Nobrega A et al (1994) Selectivity of recognition of variable (V) regions of autoantibodies by intravenous immunoglobulin (IVIg). Clin Immunol Immunopathol 70:124-128
27. Veneselli E, Conte M, Biancheri R et al (1998) Effect of steroid and high-dose immunoglobulin therapy on opsoclonus-myoclonus syndrome occurring in neuroblastoma. Med Pediatr Oncol 30:15-17
28. Dyken P, Kolar O (1968) Dancing eyes, dancing feet: infantile polymyoclonus. Brain 91:305-320
29. Bellur SN (1975) Opsoclonus: clinical value. Neurology 25:502-507
30. Rivner MH, Jay WM, Green JB, Dyken PR (1982) Opsoclonus in Hemophilus influenzae meningitis. Neurology 32:661-663
31. Kostulas V, Link H, Lefvert AK (1987) Oligoclonal IgG bands in cerebrospinal fluid, principles for demonstration and interpretation based on findings in 1114 neurological patients. Arch Neurol 44:1041-1044
32. Stephenson JBP, Graham-Pole J, Ogg L, Cochran AJ (1976) Reactivity to neuroblastoma extracts in childhood cerebellar encephalopathy. Lancet 2:975-976
33. Martin EF, Beckwith JB (1968) Lymphoid infiltrates in neuroblastoma. Their occurrence and prognostic significance. J Pediatr Surg 3:161-164
34. Ollert MW, David K, Vollmert C et al (1997) Mechanisms of in vivo anti-neuroblastoma activity of human natural IgM. Eur J Cancer 33:1942-1948
35. David K, Ehrhardt A, Ollert MW et al (1997) Expression of a 260 kDa neuroblastoma surface antigen, the target of cytotoxic natural human IgM: correlation to MYCN amplification and effects od retinoic acid. Eur J Cancer 33:1937-1941
36. Noetzel MJ, Cawley LP, James VL et al (1987) Anti-neurofilament antibodies in opsoclonus-myoclonus. J Neuroimmunol 15:137-145
37. Connolly AM, Pestronk A, Mehta S et al (1997) Serum autoantibodies in childhood opsoclonus-myoclonus syndrome: an analysis of antigenic targets in neural tissues. J Pediatr 130:878-884
38. Stefansson K, Marton LS, Dieperink ME et al (1985) Circulating autoantibodies to the 200,000-dalton protein of neurofilaments in the serum of healthy individuals. Science 228:1117-1119
39. Antunes NL, Khakoo Y, Matthay KK et al (2000) Antineuronal antibodies in patients with neuroblastoma and paraneoplastic opsoclonus-myoclonus. J Pediatr Hematol Oncol 22:315-320
40. Dalmau J, Furneaux HM, Cordon-Cardo C, Posner JB (1992) The expression of the Hu (paraneoplastic encephalomyelitis/sensory neuronopathy) antigen in human normal and tumor tissues. Am J Pathol 141:881-886
41. Anderson NE, Rosenblum MK, Graus F (1988) Autoantibodies in paraneoplastic syndromes associated with small cell lung cancer. Neurology 38:1391-1398
42. Anderson NE, Budde-Steffen C, Rosenblum MK et al (1988) Opsoclonus, myoclonus, ataxia and encephalopathy in adults with cancer: a distinct paraneoplastic syndrome. Medicine 67:100-109
43. Jensen KB, Dredge BK, Stefani G et al (2000) Nova-1 regulates neuron-specific alternative splicing and is essential for neuronal viability. Neuron 25:359-371

44. Pranzatelli MR, Huang Y, Tate E et al (1995) Cerebrospinal fluid 5-hydroxyindoleacetic acid and homovanillic acid in the pediatric opsoclonus-myoclonus syndrome. Ann Neurol 37:189-197
45. Voltz R, Dalmau J, Posner JB, Rosenfeld MR (1998) T-cell receptor analysis in anti-Hu associated paraneoplastic encephalomyelitis. Neurology 51:1146-1150
46. Ichiba N, Miyake Y, Sato K et al (1988) Mumps-induced opsoclonus-myoclonus and ataxia. Pediatr Neurol 4:224-227
47. Sheth RD, Horwitz SJ, Aronoff S et al (1995) Opsoclonus myoclonus syndrome secondary to Epstein-Barr virus infection. J Child Neurol 10:297-299
48. Delreux V, Kevers L, Callewaert A, Sindic C (1989) Opsoclonus secondary to an Epstein-Barr virus infection [letter]. Arch Neurol 46:480-481
49. Kuban KC, Ephros MA, Freeman RL et al (1983) Syndrome of opsoclonus-myoclonus caused by Coxsackie B3 infection. Ann Neurol 13(1):69-71
50. Adams GA, Shochat SJ, Smith EI et al (1993) Thoracic neuroblastoma: a pediatric oncology group study. J Pediatr Surg 28:372-378
51. Janns A, Wsladky J, Chatten J, Johnson J (1996) Opsoclonus/myoclonus: paraneoplastic syndrome of neuroblastoma. Med Pediatr Oncol 26:272-279
52. Shapiro B, Shulkin BL, Hutchinson RJ et al (1994) Locating neuroblastoma in the opsoclonus-myoclonus syndrome. J Nucl Biol Med 38:545-555

Chapter 12

Epilepsy and Autoantibodies

P. Bernasconi[1], T. Granata[2], F. Baggi[1], L. Passerini[1] and R. Mantegazza[1]

For many years the central nervous system (CNS) has been considered an immune privileged site, since alloengraftment within the CNS induced a poor immune response. This weak response was explained by: (1) the presence of a blood-brain barrier (BBB); (2) constitutive lack or low expression of proteins of the immune system, such as major histocompatibility complex (MHC) class I and II, co-stimulatory and accessory molecules, on cells of the CNS (glial cells, neurons, astrocytes); and (3) lack of lymphatic drainage [1] (see also Introduction to this volume). Nevertheless, the CNS is frequently the target of immune-mediated reactions which may have an autoimmune or an infectious aetiology. Through the years, it has become evident that autoantibodies against brain antigens exist and might lead to brain impairment and, in some disorders, are associated with seizures [2]. In this chapter, we will focus on the CNS disorders characterized by epilepsy and the presence of autoantibodies against neuronal antigens (Table 1).

Rasmussen's Encephalitis

Rasmussen's encephalitis (RE) is a rare progressive form of epilepsy that affects otherwise normal children in the first decade of life, commonly between the ages of 1 and 10 years, with no major difference in incidence between males and females [3, 4]. The disease starts with benign focal seizures that gradually develop into epilepsia partialis continua, accompanied by hemiparesis and mental retardation. Imaging data show progressive atrophy of one hemisphere [5]. The histopathological findings show perivascular cortical and meningeal T lymphocytes, a few B cells, macrophages in areas of tissue destruction, proliferation of microglial nodules, neuronal loss and gliosis [6]. Seizures are refractory to all conventional anti-epileptic drug therapies [3]. Plasmapheresis or protein A adsorption, or treatment with corticosteroids or intravenous immunoglobulins have been reported to be beneficial in some cases [7-12]. In the majority of patients

[1] Myopathology and Immunology Unit, Department of Neuromuscular Diseases; [2] Department of Child Neurology, National Neurological Institute "Carlo Besta", Via Celoria 11, 20133 Milan, Italy. e-mail: rmantegazza@istituto-besta.it

Table 1. Neurological diseases characterized by epilepsy and autoantibodies

	Antibody Target	Immuno-therapy
Rasmussen's encephalitis (RE)		
Epilepsia partialis continua	GluR3	Plasma exchange
Drug-resistant		or protein A
		immunoadsorption
Systemic lupus erythematosus (SLE)		
Primary generalized before SLE onset	Phospholipid,	Not reported
Focal or generalized-tonic during SLE	Cardiolipin	
	β_2-glycoprotein I	
Therapy-resistant localization-related		
epilepsy	Cardiolipin, nuclear,	Not reported
	β_2-glycoprotein I	
	GAD[a]	
Newly diagnosed seizure	Cardiolipin, nuclear	Not reported
	β_2-glycoprotein I	
Generalized epilepsy syndromes	Cardiolipin	Not reported
West's syndrome		Corticosteroids,
		intravenous therapy
Cryptogenic Lennox-Gastaut syndrome	Haemocyanin	Intravenous therapy
Completely-controlled epilepsy	GAD[a]	Not reported

[a] The presence of anti-GAD antibodies in uncontrolled and completely-controlled epilepsy is disputed [56, 57]

with RE, hemispherectomy remains the only procedure able to control the epileptic symptoms. Similar clinical, neuroradiological and pathological features have been described in a few cases of adult-onset severe, refractory epilepsy, suggesting that these cases may represent an adult variant of childhood RE [12-16].

RE has been proposed to be a humorally mediated autoimmune disease, based on identification of subunit 3 of the glutamate receptor (GluR3) as an antigenic target. GluR3 is a glutamate-gated ion channel belonging to the family of α-amino-3-hydroxy-5-methyl-isoxazole-4-propionic acid (AMPA)/kainate receptors known to play an important role in neuronal plasticity and neurotoxicity in the CNS [17]. In order to obtain subtype-specific antibodies against recombinant glutamate receptor subunits, four rabbits were immunized with a portion of the recombinant GluR3, fused to bacterial protein trpE, and two of them developed epileptic seizures [9]. Histopathological examination of brain sections showed perivascular lymphocytic infiltration mainly in the cerebral cortex, as well as the meninges, and microglial nodules. These features are similar to those observed in RE patients, except that the brain atrophy involved both hemispheres and not only one as in human patients [9]. On the basis of these findings, several groups looked for the presence of anti-GluR3 antibodies in sera of patients affected by RE

[7, 9, 18-20]. A variety of different approaches have been used, including ELISA with different GluR3 peptides, immunocytochemistry or immunoblotting using GluR3-transfected human embryonic kidney 293 cells, or immunohistochemistry using GluR3-affinity-purified IgG on rat brain sections. Anti-GluR3 antibodies were detected in subgroups of RE patients [7, 9, 18-20], but not in other series with well-established diagnosis of RE [12, 16], and in some patients affected by other forms of epilepsy [18, 20].

The explanation as to why autoantibodies are responsible for the progressive destruction of only one cerebral hemisphere remains a challenging problem. Two hypotheses regarding anti-GluR3 antibody action have been postulated one proposing neuron excitotoxic injury [21] and the other a mechanism based on complement-mediated cell damage [22]. Twyman and coworkers found that anti-GluR3 antibodies obtained from GluR3-immunized rabbits or RE patients cause excessive activation of the glutamate receptor, leading to destruction of the cortical neurons via an excitotoxic mechanism [21]. The excitotoxic hypothesis was further supported by the observations of Levite et al. [23] that antibodies from mice immunized with GluR3B peptide (amino acids 372-395) were able to bind to cultured neurons, evoke GluR ion channel activity and induce neuronal death; by contrast, antibodies from mice immunized with GluR3A peptide (amino acids 245-274) were able to bind to cultured neurons but failed to activate the receptor or to induce excitotoxic injury [23]. These data indicate that only a subset of antibodies can cause neuronal damage. Recently, we tested GluR3A and GluR3B affinity-purified sera from our RE patients for their capacity to induce cell damage. Both antibodies stained rat cortical neurons but were not able to evoke an ion current by patch-clamping on freshly dissociated motor neurons [19]. On the basis of our results it seems that human RE sera do not induce any inward current on cortical neurons and would not have excitotoxic properties. Additional experiments by He et al., however, demonstrated that anti-GluR3 antibodies obtained from GluR3-immunized rabbits can promote the in vitro death of cortical neurons by a complement-mediated mechanism [22]. Moreover, IgG and membrane attack complexes (MAC) were localized on the neuronal cell bodies of sections of brain cortex resected from RE patients [22, 24]. However, it is not certain that the complement fixing was induced by anti-GluR3 antibodies, and IgG and MAC localization was observed only in a subset of RE patients – those characterized by an active progressive form of RE [24]. Recently, it has been demonstrated that brain cells are capable of synthesizing complement and also express complement receptors; moreover, while astrocytes are resistant to the lytic effects of complement, oligodendrocytes and neurons are much more susceptible to killing by complement [25]. Thus, Whitney et al. hypothesized that local generation and activation of complement, triggered perhaps by unknown viral or bacterial antigens or trauma, together with seizures, might contribute to locally increased BBB permeability and unihemispheric exposure of brain antigens to pathogenic circulating antibodies. These antibodies would in turn induce neuronal cell death, thus perpetuating inflammatory reactions and, hence, seizures [24].

Very recently, it has been postulated that GluR3 is not the sole antigen in RE.

Yang et al. reported that a 9-year-old patient affected by RE was positive for anti-Munc-18 antibodies other than anti-GluR3 [26]. Munc-18 is a cytosolic protein which belongs to a protein family highly conserved from yeast to humans and involved in the exocytosis of secretory vesicles. Munc-18 is expressed in neurons, is localized in axon and presynaptic terminals, and binds several proteins such as syntaxin 1A, mint 1 and 2, and cyclin-dependent kinase 5 [27]. It is possible that anti-Munc 18 antibodies arise as part of a down-stream, or secondary, event following a humoral immune attack that destroys the synapses and leads to exposure of intracellular proteins. Thus, the role of anti-Munc 18 antibodies in the pathogenesis of RE is not clear [26], and it is more likely that they are analogous to the antibodies to other intracellular proteins, such as Yo and amphiphysin, in the paraneoplastic disorders (see Chaps. 7, 8). Studies on a larger cohort of RE patients are necessary since only one of 14 RE patients was tested positive for anti-Munc 18 antibodies.

Recently, Lousa et al. reported a patient with refractory partial and secondarily generalized status epilepticus who exhibited a dramatic response to plasma exchange [28]. The patient's serum reacted with rat CNS neurons of the cerebral cortex, hippocampus and cerebellum, but not with 293T cells expressing GluR3. Altogether, these data indicate that in RE, and in severe epilepsies characterized by clinical overlap with RE, immunological responses are implicated, but the antigens involved may be diverse. It is relevant to observe that a systematic study of the serum antibodies in different drug-resistant epilepsies is still lacking.

Systemic Lupus Erythematosus and Antiphospholipid Antibody-Associated Diseases

Epilepsy is often associated with systemic lupus erythematosus (SLE). In 5%-10% of cases the onset of seizures is several years before the clinical onset of SLE [29]. Since thrombotic events in cortical blood vessels are observed in SLE patients, seizures may be due to immune-mediated neuronal damage or may be secondary to hypertensive encephalopathy or renal failure [30]. A significant association between epilepsy and antiphospholipid antibodies has been found in SLE patients [31]. Purified IgG containing antiphospholipid antibodies has been demonstrated to depolarize synaptosomes isolated from rat brain stem [32]. Antiphospholipid antibodies may exacerbate SLE, increasing thrombotic and nonthrombotic brain injuries [33]. Antiphospholipid antibodies are heterogeneous and this may explain the different clinical manifestations observed in SLE patients [30]. In 30%-60% of patients affected by SLE and characterized by epilepsy, antibodies are mainly against cardiolipin [34, 35] and β_2-glycoprotein I [35].

Antiphospholipid antibodies have also been found in 3 out of 23 children affected by partial epileptic seizures, but not by SLE, with no evidence of focal ischaemic lesions on magnetic resonance imaging [36]. These three patients were affected by frontal lobe epilepsy and the authors hypothesized that the antibodies induced brain damage which could be pathogenic for partial epilepsy [36]. Recently, well-defined groups of patients with seizures (therapy-resistant localization-related epilepsy, generalized epilepsy syndromes, newly diagnosed seizure disorder with no anti epileptic medication) were studied for the frequency of antinuclear, anti-β_2-glycoprotein I and anticardiolipin antibodies [37]. All seizure groups had a significantly increased prevalence of IgM anticardiolipin antibodies. The frequency of positivity for antinuclear antibodies was significantly higher in newly-diagnosed seizure patients and localization-related epilepsy patients than in generalized epilepsy patients or healthy controls. Anti-β_2-glycoprotein I antibodies were found in 9% of newly diagnosed seizure or localization-related epilepsy patients. These antibodies and antiphospholipid antibodies have been demonstrated to bind to brain structures [38, 39]. The increased prevalence of autoantibodies in these patients seems to be associated with epilepsy rather than with anti-epileptic drugs, suggesting that dysregulation of the immune system is common in epilepsy [37].

West's Syndrome and Lennox-Gastaut Syndrome

Direct involvement of the immune system in other forms of intractable epilepsy, especially those affecting children, has never been proved. Patients affected by West's syndrome respond to corticosteroids better than to anti-epileptic drugs [36]. Patients affected by Lennox-Gastaut syndrome often present with cerebellar atrophy and loss of Purkinje cells, but no antibodies to Purkinje cells have been demonstrated to be present in these patients [30]. In children affected by cryptogenic Lennox-Gastaut syndrome, a humoral immune response against haemocyanin has been found [40]. Some patients affected by West's syndrome or Lennox-Gastaut syndrome respond well to intravenous Ig therapy, but this does not imply that seizures are immune-mediated. In patients affected by a variant of the Landau-Kleffner syndrome (acquired epileptic aphasia), antibodies that react against brain endothelial cells and cell nuclei have been found [41].

Epilepsy Associated with Anti-GAD Antibodies

Glutamic acid decarboxylase (GAD) catalyses the conversion of L-glutamic acid to γ-aminobutyric acid (GABA), the major inhibitory neurotransmitter in mammalian brain [42]. GAD is mainly expressed in GABA-ergic neurons and in pancreatic β cells [43]. GAD exists in different isoforms from one neurone to another [44], and it is synthesized in excess, providing a large reserve within the brain [45]. Anti-GAD antibodies are characteristic of patients affected by insulin-dependent diabetes mellitus (IDDM) and those affected by stiff-man syndrome (SMS) [46]. In IDDM patients, the antibodies are thought to be involved in the destruction of insulin-secreting cells [47], while in SMS they may inhibit GABA synthesis [48]. Impairment of GABA function has been demonstrated to provoke seizures [49], while some anti-epileptic drugs facilitate GABA-ergic activities [50]. Therefore, several groups have looked at different forms of epilepsy, in particular those forms refractory to treatment, and have found elevated levels of anti-GAD antibodies [51-55]. Further, to understand the effective role of these antibodies in inducing epilepsy, studies on a large cohort of patients with controlled or uncontrolled epilepsy were performed [56, 57]. Peltola et al. investigated the presence of anti-GAD antibodies in patients affected by therapy-resistant localization-related epilepsy and by generalized epilepsy [56]. Anti-GAD antibodies were found only in a subgroup (8 out of 51) of patients affected by localization-related epilepsy: 6 patients had a low titre, typical of IDDM, and 2 patients a high titre, similar to that seen in SMS. These 2 patients had long-standing therapy-resistant temporal lobe epilepsy and one of them had a history of autoimmune disease. These data indicate that autoimmunity against GAD might be associated with drug-resistant epilepsy. In contrast to this, Kwan et al. did not find any differences in anti-GAD antibody titres between patients with well-controlled or uncontrolled epilepsy, suggesting that anti-GAD antibodies were not associated with refractoriness to anti-epileptic therapies [57]. Further studies on cell-mediated GAD autoimmunity, on characterization of immunodominant GAD epitopes in epileptic patients, and on more homogeneous and well-defined subgroups of patients, are necessary to understand whether there are differences in GAD autoimmunity between epilepsy and the other anti-GAD-associated diseases, and whether these antibodies are causative or a secondary phenomenon.

Loss of B Cell Tolerance

The mechanisms responsible for loss of B cell tolerance in humans are still under debate (Fig. 1). Dysregulation of B cell function at different molecular levels, or alteration of the interactions between T and B lymphocytes, can result in autoimmune responses [58]. Moreover, Archelos and Hartung [58] suggested that autoimmunity can be generated by other factors such as B cell superantigens (B-SAgs) [59]. SAgs are bacterial or viral proteins which bind T or B cells via the variable (V) region of the chains outside the antigen-binding groove; on B lymphocytes the binding site for B-SAgs is represented by the V-gene segment of the heavy chain (V_H) of Ig. The activation of B lymphocytes by B-SAgs is regulated by the ratio between soluble Igs and B cells, by the isotype of the local Ig, and by the capacity of B cells to undergo somatic hypermutation [59]. In some forms of epilepsy, in particular in those where viruses have been detected, we could hypothesize that stimulation of B cells via binding of a B-SAg may contribute to amplification of immune responses within the CNS.

Another possible mechanism responsible for breakdown of B cell tolerance might be molecular mimicry [60], in which an infectious agent has a similar structure to the autoantigenic determinants of the host. The immune system at first reacts against the pathogen but, unable to discriminate between foreign and host antigens, reacts also against "self" proteins, resulting in the autoimmune

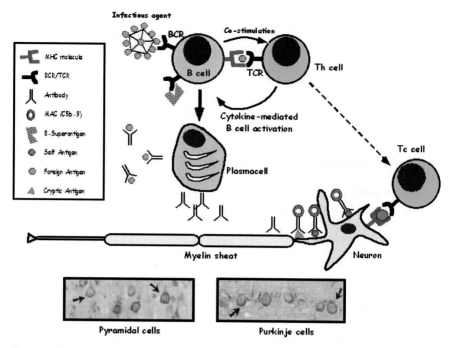

Fig. 1. Mechanisms leading to B cell activation and autoantibody production. B cells recognize an infectious agent or are activated by B-cell superantigen; both pathways may lead to pathogen-specific antibody production. If a foreign antigen (*green circle*) possess antigenic similarity to a self antigen (*red circle*) then antibodies can react also to self antigen, inducing tissue damage, usually complement-mediated (molecular mimicry). In the course of tissue alteration, cryptic antigens (*orange triangle*) may be uncovered and a process called epitope spreading may take place. Activated B cells can present the antigen via major histocompatibility complex (*MHC*) molecules to helper T lymphocytes (*Th*), which in turn activate other B cells/plasma cells to secrete antibodies or activate T cells with a cytotoxic phenotype (*Tc*); the latter cells may have a pathogenic effect on target tissue by recognition of mimicked self antigens presented by MHC molecules. In the *lower panels* an example of autoantibody reactivity is shown: immunolabelling of pyramidal and Purkinje cells (*black arrows*) was obtained using GluR3 affinity-purified human antibodies from a patient with Rasmussen's encephalitis. *BCR*, B cell receptor; *TCR*, T cell receptor; *MAC*, membrane attack complex

process. B cells, stimulated by activated T helper lymphocytes, secrete antibodies that damage the target tissue via the complement pathway. Moreover, cytotoxic T lymphocytes recognize the mimicked antigen on damaged tissue, contributing to damage [60]. This scenario (presence of antibodies, complement and CD8+ T cells) can be observed, for example, in brain sections of patients affected by RE [6, 24], and the extracellular domain of GluR3 has a structural homology with proteins expressed by bacterial strains [61]. Immune-mediated tissue damage might uncover cryptic antigens generating an antibody response against other determinants of the same molecule (intramolecular epitope spreading), or against other antigens expressed in the same target tissue (intermolecular epitope spreading) [62]. The epitope spreading phenomenon might be implicated in the anti-Munc-18 antibody response in RE patients [26], since it has been shown to occur in the anti-GAD antibody response in diabetic patients [63] and in patients affected by SLE [64].

References

1. Barker CF, Billingham RE (1977) Immunologically privileged sites. Adv Immunol 25:1-54
2. Aarli JA (1993) Immunological aspects of epilepsy. Brain Dev 15:41-50
3. Rasmussen T, Olszeweski J, Lloyd-Smith D (1958) Focal seizures due to chronic localized encephalitis. Neurology 8:435-445
4. Dulac O (1996) Rasmussen's syndrome. Curr Opin Neurol 9:75-77
5. Tampieri D, Melanson D, Ethier R (1991) Imaging of chronic encephalitis. In: Andermann F (ed) Chronic encephalitis and epilepsy: Rasmussen's syndrome. Butterworth-Heinemann, Boston, pp 47-60
6. Farrell MA, Droogan O, Secor DL et al (1995) Chronic encephalitis associated with epilepsy: immunohistochemical and ultrastructural studies. Acta Neuropathol 89:313-321
7. Antozzi C, Granata T, Aurisano N et al (1998) Long-term selective IgG immunoadsorption improves Rasmussen's encephalitis. Neurology 51:302-305
8. Andrews PI, Dichter MA, Berkovic SF et al (1996) Plasmapheresis in Rasmussen's encephalitis. Neurology 46:242-246
9. Rogers SW, Andrews PI, Gahring LC et al (1994) Autoantibodies to glutamate receptor GluR3 in Rasmussen's encephalitis. Science 265:648-651
10. Dulac O, Robain O, Chiron C et al (1991) High-dose steroid treatment of epilepsia partialis continua due to chronic focal encephalitis. In: Andermann F (ed) Chronic encephalitis and epilepsy: Rasmussen's syndrome. Butterworth-Heinemann, Boston, pp 79-110
11. Hart YM, Cortez M, Andermann F et al (1994) Medical treatment of Rasmussen's syndrome (chronic encephalitis and epilepsy): effect of high-dose steroids or immunoglobulin in 19 patients. Neurology 44:1030-1036
12. Krauss GL, Campbell ML, Roche KW et al (1996) Chronic steroid-responsive encephalitis without autoantibodies to glutamate receptor GluR3. Neurology 46:247-249
13. Gray F, Serdaru M, Baron H et al (1987) Chronic localized encephalitis (Rasmussen's) in an adult with epilepsia partialis continua. J Neurol Neurosurg Psychiatry 50:747-751
14. McLachlan RS, Girvin JP, Blume WT, Reichman H (1993) Rasmussen's chronic encephalitis in adults. Arch Neurol 50:269-274
15. Hart YM, Andermann F, Fish DR et al (1997) Chronic encephalitis and epilepsy in adults and adolescents: a variant of Rasmussen's syndrome ? Neurology 48:418-424
16. Villani F, Spreafico R, Farina L et al (2001) Positive response to immunomodulatory therapy in an adult patient with Rasmussen's encephalitis. Neurology 56:248-250
17. Monaghan DT, Bridges RJ, Cotman CW (1989) The excitatory amino acid receptors: their classes, pharmacology, and distinct properties in the function of the central nervous system. Ann Rev Pharmacol Toxicol 29:365-402
18. Mantegazza R, Bernasconi P, Aurisano N et al (2000) Frequency of anti-GluR3 anti-

bodies in Rasmussen's encephalitis, other drug-resistant epilepsies, and autoimmune diseases. Neurology 54 (Suppl.3):A246

19. Frassoni C, Spreafico R, Franceschetti S et al (2001) Labeling of rat neurons by anti GluR3 IgG from patients with Rasmussen's encephalitis. Neurology (in press)

20. Wiendl H, Bien CG, Bernasconi P et al (2001) GluR3-antibodies: prevalence in focal epilepsy but no specificity for Rasmussen's encephalitis. Neurology (in press)

21. Twyman RE, Gahring LC, Spiess J, Rogers SW (1995) Glutamate receptor antibodies activate a subset of receptors and reveal an agonist binding site. Neuron 14:755-762

22. He X-P, Patel M, Whitney KD et al (1998) Glutamate receptor GluR3 antibodies and death of cortical cells. Neuron 20:153-163

23. Levite M, Fleidervish IA, Schwarz A et al (1999) Autoantibodies to the glutamate receptor kill neurons via activation of the receptor ion channel. J Autoimmun 13:61-72

24. Whitney KD, Andrews PI, McNamara JO (1999) Immunoglobulin G and complement immunoreactivity in the cerebral cortex of patients with Rasmussen's encephalitis. Neurology 53:699-708

25. Morgan BP, Gasque P (1996) Expression of complement in the brain: role in health and disease. Immunol Today 17:461-466

26. Yang R, Puranam RS, Butler LS et al (2000) Autoimmunity to Munc-18 in Rasmussen's encephalitis. Neuron 28:375-383

27. Misura KM, Scheller RH, Weis WI (2000) Three-dimensional structure of the neuronal-Sec-1-syntaxin 1a complex. Nature 404:355-362

28. Lousa M, Sanchez-Alonso S, Rodriguez-Diaz R et al (2000) Status epilepticus with neuron-reactive serum antibodies: response to plasma exchange. Neurology 54:2163-2165

29. Mackworth-Young CG, Hughes GR (1985) Epilepsy: an early symptom of systemic lupus erythematosus. J Neurol Neurosurg Psychiatry 48:185-192

30. Aarli JA (2000) Epilepsy and the immune system. Arch Neurol 57:1689-1692

31. Herranz MT, Rivier G, Khamashta MA et al (1994) Association between antiphospholipid antibodies and epilepsy in patients with systemic lupus erythematosus. Arthritis Rheum 37:568-571

32. Chapman J, Cohen-Armon M, Shoenfeld Y, Korczyn AD (1999) Antiphospholipid antibodies permeabilize and depolarise brain synaptoneurosomes. Lupus 8:127-133

33. Sabet-Arman SWL, Stidley CA, Danska J, Brooks WM (1998) Neurometabolite markers of cerebral injury in the antiphospholipid antibody syndrome of systemic lupus erythematosus. Stroke 29:2254-2260

34. Liou HH, Wang CR, Chen CJ et al (1996) Elevated levels of anticardiolipin antibodies and epilepsy in lupus patients. Lupus 5:307-312

35. Shrivastava A, Dwivedi S, Aggarwal A, Misra R (2001) Anti-cardiolipin and anti-β_2 glycoprotein I antibodies in Indian patients with systemic lupus erythematosus: association with the presence of seizures. Lupus 10:45-50

36. Angelini L, Granata T, Zibordi F et al (1998) Partial seizures associated with antiphospholipid antibodies in childhood. Neuropediatrics 29:249-253

37. Peltola JT, Haapala A, Isojärvi JI et al (2000) Antiphospholipid and antinuclear antibodies in patients with epilepsy or new-onset seizure disorders. Am J Med 109:712-717

38. Caronti B, Pittoni V, Palladini G, Valesini G (1998) Anti-β_2-glycoprotein I antibodies bind to central nervous system. J Neurol Sci 156:211-219

39. Kent M, Alvarez F, Bogt E et al (1997) Monoclonal antiphosphatidylserine antibodies react directly with feline and murine central nervous system. J Rheumatol 24:1725-1733

40. van Engelen BG, Weemaes CM, Renier WO et al (1995) A dysbalanced immune system in cryptogenic Lennox-Gastaut syndrome. Scand J Immunol 41:209-213

41. Connolly AM, Chez MG, Pestronk A et al (1999) Serum autoantibodies to brain in Landau-Kleffner variant autism and other neurologic disorders. J Pediatr 134:607-613

42. Erlander MG, Tobin AJ (1991) The structural and functional heterogeneity of glutamic acid decarboxylase: a review. Neurochem Res 16:215-226

43. Erdo S, Wolff J (1990) Gamma-aminobutyric acid outside the mammalian brain. J Neurochem 54:363-372

44. Solimena M, Butler MH, De Camilli P (1994) GAD, diabetes, and stiff-man syndrome: some progress and more questions. J Endocrinol Invest 17:509-520

45. Martin DL, Rimvall K (1993) Regulation of γ-aminobutyric acid synthesis in the brain. J Neurochem 60:395-407

46. Solimena M, Folli F, Aparisi R et al (1990) Autoantibodies to GABA-ergic neurons and pancreatic beta cells in stiff-man syndrome. N Engl J Med 322:1555-1560

47. Lohmann T, Hawa M, Leslie RDG et al (2000) Immune reactivity to glutamic acid decarboxylase 65 in stiff-man syndrome and type 1 diabetes mellitus. Lancet 356:31-35

48. Dinkel K, Meinck H-M, Jury KM et al (1998) Inhibition of γ-aminobutyric acid synthesis by glutamic acid decarboxylase autoantibodies in stiff-man syndrome. Ann Neurol 44:194-201

49. Meldrum BS (1995) Neurotransmission in epilepsy. Epilepsia 36:S30-S35

50. Macdonald RL, Kelly KM (1995) Antiepileptic drug mechanisms of action. Epilepsia 36:S2-S12

51. Solimena M, Folli F, Denis-Donini S et al (1988) Autoantibodies to glutamic acid decarboxylase in a patient with stiff-man syndrome, epilepsy, and type 1 diabetes mellitus. N Engl J Med 318:1012-1020

52. Saiz A, Arpa J, Sagasta A et al (1997) Autoantibodies to glutamic acid decarboxylase in three patients with cerebellar ataxia, late-onset insulin-dependent diabetes mellitus, and polyendocrine autoimmunity. Neurology 49:1026-1030

53. Nemni R, Braghi S, Natali-Sora MG et al (1994) Autoantibodies to glutamic acid decarboxylase in palatal myoclonus and epilepsy. Ann Neurol 36:665-667

54. Giometto B, Nicolao P, Macucci M et al (1998) Temporal-lobe epilepsy associated with glutamic-acid-decarboxylase autoantibodies. Lancet 352:457

55. Marchiori GC, Vaglia A, Vianello M et al (2001) Encephalitis associated with glutamic acid decarboxylase autoantibodies. Neurology 56:814

56. Peltola J, Kulmala P, Isojärvi J et al (2000) Autoantibodies to glutamic acid decarboxylase in patients with therapy-resistant epilepsy. Neurology 55:46-50

57. Kwan P, Sills GJ, Kelly K et al (2000) Glutamic acid decarboxylase autoantibodies in controlled and uncontrolled epilepsy: a pilot study. Epilepsy Res 42:191-195

58. Archelos JJ, Hartung H-P (2000) Pathogenetic role of autoantibodies in neurological diseases. Trends Neurosci 23:317-327

59. Silverman GJ (1997) B-cell superantigens. Immunol Today 18:379-386

60. Albert LJ, Inman RD (1999) Molecular mimicry and autoimmunity. N Engl J Med 341:2068-2074

61. O'Hara PJ, Sheppard PO, Thogersen H et al (1993) The ligand-binding domain in metabotropic glutamate receptors is related to bacterial periplasmic binding domains. Neuron 11:41-52

62. Sercarz EE, Lehmann PV, Ametani A et al (1993) Dominance and crypticity of T cell antigenic determinants. Ann Rev Immunol 11:729-766

63. Sohnlein P, Muller M, Syren K et al (2000) Epitope spreading and a varying but not disease-specific GAD65 antibody response in type I diabetes. The Childhood Diabetes in Finland Study Group. Diabetologia 43:210-217

64. Deshmukh US, Lewis JE, Gaskin F et al (2000) Ro60 peptides induce antibodies to similar epitopes shared among lupus-related autoantigens. J Immunol 164:6655-6661

Chapter 13

Primary and Secondary Vasculitis of the Central Nervous System

V. Martinelli[1], A. Manfredi[2], L. Moiola[1], M.G. Sabbadini[2], G. Comi[1]

Vasculitis is characterised by inflammation of the arterial wall, with possible involvement of adjacent tissues. Besides tissue injury due to ischaemia, sustained inflammation sometimes compromises vessel integrity. The central nervous system (CNS) is the target organ of isolated, or primary, vasculitis (primary angiitis of the CNS, or PACNS). Far more frequently, the involvement of the CNS is part of a more complex scenario involving other organs or systems (systemic vasculitis)[1]. Understanding the immunopathological processes underlying vasculitis, as well as the role of both local and distal regulatory control on vascular inflammation, remains an intriguing field of interest.

Immunopathogenetic Mechanisms

Initiation and Maintenance of Vascular Inflammation

Agents that interfere with the homoeostatic function of endothelia cause inflammation. In most conditions, the inflammatory reaction is *transient*. The reason is possibly two-fold: (1) the release of pro-inflammatory factors and the up-regulation of adhesion molecules are temporally limited, and (2) anti-inflammatory signals are generated actively in parallel. This is the case for the acute syndromes due to chemical agents, which directly harm the endothelial lining. More frequently, infections by bacteria, viruses, fungi and protozoa cause clinically relevant vasculitis, often involving the nervous system.

Persistent inflammation is caused by locally expressed antigens (usually infectious), non-healing lesions of the vessel wall with exposure of matrix components, or maintained autoimmune responses. Persistent inflammation can be due to a bona fide immune response targeting components of the vessel, as with anti-basement membrane antibodies in Goodpasture's syndrome. More frequently, inflammatory processes not originally directed towards the vasculature eventually involve it. This occurs, for example, in hypersensitivity diseases with immune

[1] Department of Neuroscience; [2] Department of Medicine, San Raffaele Scientific Institute, Via Olgettina 48, 20132 Milan, Italy. e-mail: vittorio.martinelli@hsr.it

complex vasculitis. Immune complexes with defined chemical components form at the endothelial surface or are trapped in situ. They activate the complement cascade and cause local activation of polymorphonuclear leucocytes, with recruitment of cytotoxic activities and release of pro-inflammatory mediators. In the case of diseases characterised by the preferential inflammation of large arteries (see below), the unrestrained activation of T cells leads to a granulomatous reaction with tissue destruction, including necrosis of smooth muscle cells and fragmentation of elastic membranes. Inflammation often progresses to a sclerotic stage, with intimal and adventitial fibrosis and scarring of the media.

The association of central and peripheral nervous system vasculitis with neoplasia seldom occurs, with the notable exception of angiotrophic lymphoma. Neoplastic cells are in this case localised in the vasculature and the clinical presentation resembles that of systemic necrotising vasculitis, often associated with CNS involvement [2]. Systemic vasculitis also develops in association with systemic autoimmune diseases. However, with the possible exception of systemic lupus erythematosus, with either overt or subclinical CNS involvement [3], vasculitis of the CNS is not typical.

Mechanisms Involved in Tissue Damage During Vasculitis

All the factors outlined above disturb the function of the vascular endothelium, which is specialised in the control of leucocyte recirculation, in the maintenance of resistance to thrombosis and in the regulation of vascular tone. Infiltration of the vessel wall by inflammatory cells is the hallmark of vascular inflammation. Multiple receptor-ligand interactions are involved in the capture of leucocytes from the blood, in their firm adhesion to endothelia and in their migration through the vessel wall. The principal interactions occur among selectins, integrins and receptors belonging to the immunoglobulin superfamily. Pro-inflammatory factors up-regulate the expression of adhesion molecules and enhance the efficiency of their reciprocal interaction. Specific chemoattractants recruiting different leucocyte subpopulations, further shape the characteristics of the vessel inflammation.

Direct involvement of the vessel wall and vasoconstriction impede the blood flow to peripheral tissues, which become insufficient for tissue metabolic demands. The release of endothelin further increases the vascular tone, worsening ischaemia. This feature should be kept in mind when interpreting the results of cerebral angiography. Furthermore, during inflammation, the balance between procoagulant and anticoagulant properties associated with the vascular endothelium shift towards a procoagulant effect [4].

A Role for Autoantibodies in the Initiation or Maintenance of Vasculitis

Antineutrophil cytoplasmic antibodies (ANCA) comprise a heterogeneous group of auto-antibodies, which are routinely detected by indirect immunofluorescence. At least three patterns can be recognised: the classic granular cytoplasmic pattern (cANCA), the perinuclear fluorescence pattern (pANCA) and an atypical, more diffuse pattern (aANCA). Proteinase 3 is the antigen to which most antibodies bind to yield the classic ANCA pattern. These antibodies are highly sensitive and specific for systemic necrotising Wegener's granulomatosis (see below). By contrast, pANCA directed against another azurophilic granule antigen, myeloperoxidase, associate more frequently with microscopic polyangiitis. Direct evidence of a role for ANCA in causing vasculitis is currently lacking: transplacental transfer of ANCA is not associated with transfer of vasculitis during pregnancy, and the presence of ANCA in intravenous immunoglobulin preparations seems insufficient to cause any symptoms in the recipients. In accordance with this, administration of ANCA into baboons has never caused vasculitis. Furthermore, antibodies are absent from the vasculitic lesions.

However, other evidence does implicate ANCA in the pathogenesis of small-vessel vasculitis [5]: titres correlate with the clinical activity of the disease, and it is possible that their elevation helps to predict relapses of the disease. In vitro, cytokine activation (priming) of neutrophils and monocytes causes the redistribution of granule antigens, including those recognised by ANCA, to the plasma membrane. Purified ANCA antibodies bind to primed neutrophils and increase their expression of integrins, facilitating their adherence to endothelia and their transmigration to subendothelial spaces. In parallel, after cross-linking of membrane antigens by ANCA, activated neutrophils are able to kill endothelial cells efficiently. Of interest, ANCA protects proteinase 3 from the inactivation caused by α_1-antitrypsin, possibly magnifying the ability of the neutral protease to cause tissue damage. Therefore, although there is no evidence that ANCA can *initiate* the disease, the antibodies may represent formidable *enhancers* of the inflammatory response, amplifying the effect of cytokines on activated leucocytes and disturbing both the interactions between endothelia and inflammatory cells and the regulatory mechanisms that limit the proteolytic activity of released neutrophil enzymes.

Recently, ANCA have been demonstrated to be endowed with the ability to opsonise dying granulocytes, an event that leads to their enhanced phagocytic clearance in a pro-inflammatory environment [6]. Opsonised apoptotic cells represent a preferential source of antigens for antigen presenting cells, leading to the efficient activation of T cells specific for intracellular antigens of the dying cells [7, 8]. Therefore, opsonisation of apoptotic neutrophils may represent a self-perpetuating event involved in the maintenance of the autoimmune response against neutrophil antigens.

CNS a Privileged Site Protected Against Vasculitic Injury

Reports of an association between ANCA-associated small vessel vasculitis (in particular Wegener's granulomatosis and microscopic polyangiitis) and CNS involvement are not uncommon. In at least one patient with Wegener's granulomatosis, repeated intrathecal administration of methotrexate and steroids was effective in arresting the progression of meningeal infiltration. This was associated with the disappearance of cANCA from the cerebrospinal fluid [9]. However, the CNS per se is relatively spared, in sharp contrast to the involvement of the peripheral nervous system (PNS), a prominent feature of systemic and secondary vasculitis. Accordingly, while degenerative or thrombotic features are detected, inflammation of CNS vessels is unusual, suggesting the existence at this site of mechanisms minimising vascular inflammation that are not present in the PNS [4]. These mechanisms are apparently, in most cases, sufficient to bypass the facilitator action of auto-antibodies (see above).

Moore and Richardson [4] propose that at least three different mechanisms are involved: diminished signalling, reduced leucocyte trafficking and tighter regulation over inflammatory responses. The hypothesis of diminished signalling stems from the identification of peculiar physical and biochemical characteristics of endothelial monolayers in the CNS vasculature, with tight intercellular junctions, relatively devoid of micropinocytic particles and abundantly expressing the γ-glutamyltranspeptidase enzyme. In contrast, the limited expression of adhesion molecules possibly contributes to the reduced adhesion of circulating lymphocytes to the CNS vessels, with reduced transmigration to the parenchyma. Upon transmigration, the relative lack of major histocompatibility complex expression contributes to quench the local perpetuation of the auto-antigen-driven specific immune response. Furthermore, anti-inflammatory molecules, like TGF-β, which control both the innate and the acquired immune responses, have been suggested to play a more prominent role in the CNS than in the systemic vasculature.

These mechanisms per se, however, fail to explain why the inflammatory response, once initiated, selectively fails to persist in the CNS. In these conditions, indeed, endothelial monolayers become activated and permeable and immune cells efficiently transmigrate through the vessel walls. This is required to ensure surveillance against the potentially dangerous noxa that initiated the inflammatory response, and in particular against infectious agents. Better molecular identification of the protective mechanism operating in the CNS during systemic vasculitis may prove helpful in different clinical settings in which chronic inflammation of the CNS contributes to the clinical picture.

Classification of Vasculitis Associated with CNS Involvement

The classification of vasculitis is mainly based on clinical, histopathological, immunopathogenetic features such as the patient's age at disease onset, the preferred sites of vascular involvement, the organs involved, the clinical course of the symptoms, the characteristics of the local inflammatory process and the presence or absence of systemic factors suggesting a peripheral activation of the immune system [10, 11].

Although different grouping criteria can be used, and a universally accepted classification has not emerged [12-14], we think it is useful in this review to differentiate between primary vasculitis, secondary vasculitis and systemic vasculitis associated with connective tissue diseases or with other diseases of the immune system (Table 1).

Although a histological evaluation of brain biopsy is needed for the diagnosis of many vasculitides, the histological characteristics of this procedure are not generally specific for the different forms of vasculitis. Moreover, the size of the vessels involved is not a rigorous criterion that allows identification of the different clinical aspects of vasculitis [15]. CNS involvement, a mandatory criterion in isolated angiitis of the CNS, is extremely variable in all of the other forms of systemic vasculitis: it can be rare or unusual in some forms, much more frequent in others (Table 2).

The onset and the clinical features of vasculitis are extremely variable and therefore an early diagnosis is often difficult, especially when CNS involvement represents the only clinical expression of the underlying immune-pathological process. Knowledge of the main diagnostic criteria of the different forms of vasculitis is a prerequisite to making a correct diagnosis and starting an early treatment. Here we will briefly describe the clinical, imaging and laboratory features typical of the main vasculitides affecting the CNS, particularly isolated angiitis of CNS.

Table 1. Classification of vasculitis

Primary vasculitis
Polyarteritis nodosa (PAN)
Allergic granulomatous angiitis (Churg-Strauss vasculitis)
Temporal arteritis
Takayasu's arteritis
Hypersensitivity vasculitis
Wegener's granulomatosis
Behçet's disease
CNS isolated angiitis (PACNS, Cogan's syndrome, Susac's syndrome, spinal cord isolated vasculitis)
Lymphomatoid granulomatosis

Secondary vasculitis
Vasculitis secondary to infection (rickettsial, bacterial, fungal, viral, mycoplasmal, protozoal)
Vasculitis secondary to neoplasia (Hodgkin's lymphoma, non-Hodgkin's lymphoma, leukaemia, small cell lung cancer)
Vasculitis secondary to toxins or drugs (allopurinol, amphetamines, sympathocomimetics, cocaine, heroin)

Systemic vasculitis associated with connective tissue or immune system disease
Rheumatoid arthritis
Sjögren's syndrome
Mixed connective tissue disease
Systemic lupus erythematosus
Inflammatory intestinal diseases
Antiphospholipid syndrome

Table 2. CNS involvement in the vasculitic syndromes

Vasculitic syndrome	Neurological symptoms and signs	Incidence of neurological abnormalities (%)
Polyarteritis nodosa	Focal signs, epilepsy, headache, subarachnoid hemorrhage	23-53
Wegener's granulomatosis	Focal signs, aseptic meningitis	20-50
Temporal arteritis	Focal signs, amaurosis, headache	20-60
Churg-Strauss granulomatosis	Focal signs, encephalopathy, headache	50-60
Takayasu's arteritis	Focal signs, syncope, epilepsy	10-36
Hypersensitive vasculitis	Focal signs, epilepsy	10
CNS isolated vasculitis	Focal signs, headache	100
Rheumatoid arthritis	Focal signs, epilepsy, encephalopathy	1
Mixed connective tissue disease	Headache, encephalopathy, epilepsy	20-50
Sjögren's syndrome	Focal signs, myelopathy, epilepsy	5-15
Systemic lupus erythematosus	Focal signs, psychosis, headache, epilepsy	20-50
Behçet's disease	Meningoencephalitis, psychiatric disturbances, focal signs, headache	20-40

Polyarteritis Nodosa

Polyarteritis nodosa (PAN) is a systemic necrotising vasculitis of small- and medium-sized muscular arteries, without involvement of lung and spleen [16]. It is initially characterised by infiltration of polymorphonuclear leucocytes followed by mononuclear cells, proliferation of the intima, fibrinoid necrosis, thrombosis and ischaemia. Hypertension is very frequent because of kidney involvement. It is often associated with systemic inflammatory indices [elevation of erythrocyte sedimentation rate (ESR), antinuclear antibodies, low rheumatoid factor titre and leucocytosis] and in about 30% of cases with hepatitis B. Involvement of the PNS is much more frequent than CNS involvement and the diagnosis is sometimes based on the identification of vasculitis at nerve biopsy. Two common presentations of CNS involvement are diffuse encephalopathy and focal or multifocal disturbances. Evidence of alterations of visceral angiography is extremely typical of this vasculitis. Therapy is based on combined treatment with cyclophosphamide and prednisone.

Allergic Granulomatous Angiitis (Churg-Strauss Vasculitis)

This condition is a systemic granulomatous and necrotising vasculitis with infiltration of eosinophils, initially characterised by asthma and allergic rhinitis followed by hypereosinophilia (> 10%) with eosinophilic infiltration of lung and gastrointestinal tract (the latter phase can last 2-20 years) [17]. Purpura and subcutaneous nodules are present in approximately two-thirds of the patients. It is similar to PAN but, in addition to small- and medium-sized muscular arteries, capillaries, veins and venules can be involved and in fact it is called the vasculitis of small vessels. Constitutional symptoms are fever and weight loss; 70% of patients are positive for ANCA. The development of mononeuropathy, multiple mononeuropathies or polyneuropathies is frequent. Due to the possible involvement of myocardium (coronary arteritis and myocarditis), it is really important to make an early diagnosis in order to start therapy with high-dosage steroids and immunosuppressive agents, especially cyclophosphamide [18].

Temporal Arteritis

Temporal arteritis is a granulomatous vasculitis of medium-sized and large arteries, characteristically involving, but not limited to, one or more branches of the carotid artery, especially the temporal artery [19]. It is a giant cell arteritis occurring mostly in elderly patients (over the age of 55 years, rarely in younger patients). Typical symptoms include non-throbbing headache, temporal tenderness, jaw claudication and visual dimming, and they are associated with marked elevation of ESR (although cases with normal ESR have been described), low-grade fever, fatigue, malaise, weight loss and proximal myalgia (temporal arteritis is often associated with polymyalgia rheumatica). Diagnosis is confirmed by temporal artery biopsy, although a falsely negative biopsy may occur due to sampling error. Therapy with high-dosage steroids has to be started immediately if typical symptoms are present (amaurosis, jaw claudication, headache, temporal tenderness); short-term exposure to therapy should not greatly diminish the sensitivity of the biopsy, which should, nevertheless, be done as soon as possible. Without treatment, secondary eye involvement occurs in up to 65% of the cases, usually within 1-10 days of the initial visual loss. The disease is very responsive to prednisone and treatment has to be maintained at low dosage for 1-2 years.

Takayasu's Arteritis

Takayasu's arteritis (aortic arch syndrome, pulseless disease) is a panvasculitis of large elastic arteries with a mononuclear cells infiltrate, but giant cells are not as common as in temporal arteritis. It affects large arteries including the aorta and the major vessels arising from the aortic arch. It is a rare disease affecting young women, especially oriental women. The illness begins with an inflammatory phase with non-specific systemic symptoms and signs, followed by a chronic phase characterised by symptoms due to occlusion or stenosis of the vascular lumen. In the acute phase, anaemia and elevated ESR are common. The therapeutic response to steroid treatment is variable. Sometimes immunosuppressive, antiplatelet or anticoagulant agents have to be added to steroid therapy.

Hypersensitivity Vasculitis

Hypersensitivity vasculitis is a necrotising vasculitis of small vessels (arterioles, capillaries and venules), secondary to exposure to drugs or infections (25% of cases), but the term is often used to define the small vessel vasculitides associated with cryoglobulinaemia, malignancies, sarcoidosis, Henoch-Schönlein purpura or connective tissue diseases (50% of cases). In the remaining 25% of cases there is no identifiable cause. An infiltrate of polymorphonuclear leucocytes and fibrinoid necrosis of vessel walls is characteristic. It commonly affects the skin, producing a non-blanching, palpable, purpuric rash, but all the organs can be involved, including CNS and PNS. ESR is typically elevated and hypereosinophilia is common. The response to steroid treatment is usually good and the dose has to tapered slowly depending on disease severity and evolution.

Wegener's Granulomatosis

Wegener's granulomatosis is a necrotising, granulomatous vasculitis involving the upper respiratory tract and lungs (90% of patients), affecting small- and medium-sized vessels, with associated fever and malaise [20]. Kidney involvement is common (focal glomerulonephritis). The CNS involvement may be caused by contiguous extension of the granulomatous lesions from the sinuses to the orbits or to the base of the cranium by vasculitis involving CNS structures or, more frequently, cranial or peripheral nerves. Cranial neuropathies may be the initial symptom of the disease, are sudden in onset and may be painless. Laboratory findings include an elevated ESR, haematuria, proteinuria, hypergammaglobulinaemia and renal insufficiency. Antineutrophil cytoplasmic auto-antibodies with a cytoplasmic staining pattern (cANCA) are relatively specific for Wegener's granulomatosis (90% of cases) [21]. The classic triad of respiratory tract granulomatosis, necrotising glomerulonephritis and systemic small vessel vasculitis readily suggests the diagnosis. Treatment is with steroids and immunosuppressive agents.

Behçet's Disease

Behçet's disease is an idiopathic, multisystemic inflammatory disorder characterised by recurrent ocular lesions, oral and genital ulcers and erythema nodosum. In addition, many other organs can present small vessel vasculitis with a monocellular infiltrate without necrosis. Up to a third of patients have neurological symptoms or signs, typically of the CNS, with headache in 80% of cases, diffuse encephalopathy, aseptic meningoencephalomyelitis with fever, venous sinus thrombosis or focal signs due to ischaemic parenchymal lesions [22]. CSF findings are characterised by pleocytosis, elevation of protein and gammaglobulin levels and the presence of oligoclonal bands. The course is often benign, but neurological involvement worsens the prognosis. Therapy is based on the use of corticosteroids combined with azathioprine, cyclophosphamide, chlorambucil or cyclosporin B.

Primary Angiitis of the CNS

PACNS is an uncommon form of vasculitis, usually not associated with any change in systemic inflammatory indices, immune complexes or auto-antibodies [23].

The involvement is restricted to CNS parenchymal and leptomeningeal vessels. In the past, the disorder was called "granulomatous angiitis of the CNS" (GANS) because of its histological features: the involvement of small- and medium-sized arteries (more rarely venules), with mononuclear infiltrates, granulomata and, sometimes, multinucleated giant cells or lymphocytes [24]. Damage may be segmental and perivascular infiltration may be found in multiple sections. In contradistinction to temporal arteritis, the inflammation relatively spares the media and intima. Since histological findings suggesting a necrotising lymphocytic vasculitis have been recently described, it is now preferable to use the term "primary angiitis of the CNS" (PACNS), which underlines the absence of other factors that might induce vasculitis (drugs, toxins, infections, systemic diseases). Clinical features are extremely variable, ranging from focal neurological deficits due to ischaemic damage, or focal signs suggesting a mass lesion, to disturbances related to diffuse encephalopathy (abnormal behaviour, psychosis, impaired cognition or confusion). The earliest manifestation of PACNS is generally headache, but rarely the disease can present as a myelopathy. Definite diagnosis is based on leptomeningeal-cortical brain biopsy to confirm the presence of inflammation and to exclude other vascular or non-inflammatory diseases [25, 26].

Cerebral angiography shows abnormalities in 60% of histologically proven cases; typical angiographic abnormalities are mass effect or multiple areas of stenosis, occlusion of vessels, delayed emptying of vessels and anastomotic channels. The cerebrospinal fluid analysis can show non-specific abnormalities such as mild pleocytosis and/or elevation of protein concentration. In the past PACNS was rapidly progressive and fatal; today the course of the disease has been changed because of the introduction of diagnostic tools allowing early diagnosis and consequently early treatment with high doses of steroids and cyclophosphamide.

Accurate interpretation and use of the diagnostic techniques have recently allowed recognition of a few cases of cerebral vasculitis considered benign or minimally affected [27]. The diagnosis of these forms is not based on a brain biopsy, but only on clinical features and on the abnormalities seen by brain MRI and cerebral angiography, after the exclusion of any factor or situation that might induce similar findings. The term "benign angiopathy of the CNS" (BACNS) should be reserved for these forms of "probable" cerebral vasculitis defined only on the basis of "nonspecific" angiographical findings and with typical clinical features. Usually in BACNS there is a focal onset, normal or near-normal cerebrospinal fluid and a benign course even without therapy. Steroid treatment is the usual therapy; rarely, a combined treatment of steroids and cyclophosphamide is necessary (some authors suggest a combined treatment of steroids and calcium antagonists for 3-6 weeks).

A percentage of patients (unknown because reliable reports do not exist) are affected by PACNS but have normal cerebral angiography, since selective involvement of only small arteries and arterioles is possible. We observed six cases (two of them with leptomeningeal-cortical confirmation of small vessel vasculitis, see Fig. 1a, b) who initially had a diagnosis of multiple sclerosis and then of PACNS with negative angiography. General laboratory parameters and acute phase reac-

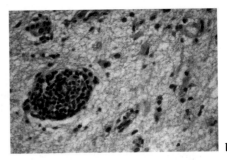

a   b

Fig. 1. a Primary angiitis CNS. Cerebral biopsy specimen showing white matter inflammatory cell infiltrate and necrosis (hermatoxylin and eosin, X 62.5. **b** Gliosis and small vessel with total obliteration of vascular wall and lumen by mononuclear cells (hematoxylin and eosin, X 125)

tants were normal in all patients. Oligoclonal bands were positive in two patients; one or more evoked potentials were abnormal in five patients. Brain MRI showed in all patients multiple abnormal areas, most of which were enhancinged after Gd-DTPA injection. The finding of intense and simultaneous enhancement of the majority of brain MRI abnormalities during a clinically active phase of the disease, not at disease presentation, can be proposed as the most important noninvasive criterion for a diagnosis of small vessel PACNS. Acute steroid treatment at high dosage, followed by tapering of the dose, improved the clinical outcome in all our patients and was associated with a dramatic reduction of the number and extent of enhancing lesions. The early discontinuation of steroid treatment caused clinical worsening and the reappearance of multiple enhancing areas in the brain, suggesting that this pathological process is sensitive to steroids. Azathioprine was added to treat four patients adequately. The optimum therapy duration is unknown, but it should be continued for at least 12 months after clinical remission has been achieved.

Finally, other forms of isolated vasculitis of CNS have been described, such as Cogan's syndrome and Susac's syndrome. The former is characterised by CNS vasculitis, interstitial keratitis and otovestibular system involvement with consequent vertigo and severe bilateral deafness. The second is a rare idiopathic vasculitis selectively involving brain, retinal vessels and cochlea, predominantly affecting young women.

Conclusions

The approach to a patient affected by a vasculitis of the CNS requires the following: (1) recognition of the inflammatory vascular pathology affecting the CNS, (2) a specific and correct diagnosis, (3) assessment of the prognosis and (4) early, specific therapy. The final goal should be to tailor the treatment taking into account the course and prognosis of the particular form of vasculitis.

References

1. Moore PM, Cupps TR (1983) Neurological complications of vasculitis. Ann Neurol 14:155-167
2. Roux S, Grossin M, De Bandt M et al (1995) Angiotropic large cell lymphoma with mononeuritis multiplex mimicking systemic vasculitis. J Neurol Neurosurg Psychiatry 58:363-366
3. Sabbadini MG, Manfredi AA, Bozzolo E et al et al (1999) Central nervous system involvement in systemic lupus erythematosus patients without overt neuropsychiatric manifestations. Lupus 8:11-19
4. Moore PM, Richardson B (1998) Neurology of the vasculitides and connective tissue diseases. J Neurol Neurosurg Psychiatry 65:10-22
5. Gross WL, Trabandt A, Reinhold-Keller E (2000) Diagnosis and evaluation of vasculitis. Rheumatology 39:245-252
6. Moosig F, Csernok E, Kumanovics G, Gross WL (2000) Opsonization of apoptotic neutrophils by anti-neutrophil cytoplasmic antibodies (ANCA) leads to enhanced uptake by macrophages and increased released of tumor necrosis factor-α. Clin Exp Immunol 122:499-503
7. Rovere P, Sabbadini MG, Vallinoto C et al (1999) Dendritic cell presentation of antigens from apoptotic cells in a proinflammatory context: role of opsonizing anti-β2 glycoprotein I monoclonal antibodies. Arthritis Rheum 42:1412-1420
8. Rovere P, Sabbadini MG, Fazzini F et al (2000) Remnants of suicidal cells fostering systemic autoaggression: apoptosis in the origin and maintenance of autoimmunity. Arthritis Rheum 43:1663-1672
9. Spranger M, Schwab S, Meinck HM et al (1997) Meningeal involvement in Wegener's granulomatosis confirmed and monitored by positive antineutrophil cytoplasm in cerebrospinal fluid. Neurology 48:263-265
10. Sigal LH (1987) The neurologic presentation of vasculitic and rheumatologic syndromes. Medicine 66:157-168
11. Chu CT, Gray L, Goldstein LB et al (1998) Diagnosis of intracranial vasculitis: a multidisciplinary approach. J Neuropathol Exp Neurol 57:30-38
12. Fieschi C, Rasura M, Anzini A et al (1998) Central nervous system vasculitis. J Neurol Sci 153:159-171
13. Moore PM (1998) Central nervous system vasculits. Curr Opin Neurology 11:241-246
14. Jennette JC, Falk RJ (1997) Small-vessel vasculitis. N Engl J Med 337:1512-1523
15. Lie JT (1996) Angiitis of the central nervous system. In: Ansell BM, Bacon PA (eds) The vasculitides. Chapman and Hall, London, pp 246-263
16. Guillevin L, Le TH, Godeau P et al (1988) Clinical findings and prognosis of polyarteritis nodosa and Churg Strauss angiitis: a study in 165 patients. Br J Rheumatol 27:258-264
17. Guillevin L, Cohen P Gayraud M et al (1999) Churg-Strauss syndrome. Clinical study and long term follow-up of 96 patients. Medicine 78:26-37
18. Sehgal M, Swanson JW, DeRemee RA, Colby TV (1995) Neurologic manifestations of Churg-Strauss syndrome Mayo Clin Proc 70:337-341
19. Keltner JL (1982) Giant-cell arteritis: signs and symptoms Ophthalmology 89:1101-1110
20. Nishino H, Rubino FA et al (1993) Neurological involvement in Wegener granulomatosis: an analysis of 324 consecutive patients at the Mayo Clinic. Ann Neurol 33:4-9
21. Gross WL, Csernok E, Flesch BK (1993) "Classic" anti-neutrophil cytoplasmic autoan-

tibodies, "Wegener's autoantigen" and their immunopathogenetic role in Wegener granulomatosis. J Autoimmun 6:171-184

22. Serdaroglu P, Yazici H et al (1989) Neurologic involvement in Behçet's syndrome: a prospective study. Arch Neurol 46:270-273

23. Vollmer TL, Guarnaccia J, Harrington W et al (1993) Idiophatic granulomatous angiitis of the central nervous system. Arch Neurol 50:925-930

24. Rhodes RH, Meadelaire NC, Petrelli M et al (1995) Primary angiitis and angiopathy of the central nervous system and their relationship to systemic giant cell arteritis. Arch Pathol Lab Med 119:334-339

25. Moore PM (1989) Diagnosis and management of isolated angiitis of the central nervous system. Neurology 39:167-173

26. Calabrese LH, Duna GF, Lie JT (1997) Vasculitis in the central nervous system. Arthritis Rheum 40:1189-1201

27. Woolfenden AR, Tong DC, Marks MP et al (1998) Angiographically defined primary angiitis of the CNS: is it really benign? Neurology 51:183-188

Subject Index